Stapleton's Real Estate Management Practice

(Fourth Edition)

Edited by
Anthony Banfield FRICS, DipProjMan
Director of Corporate Training, College of Estate Management

2005

A division of Reed Business Information

Estates Gazette
1 Procter Street, London WC1V 6EU

First edition 1981
Reprinted 1983
Second edition 1986
Third Edition 1994
Fourth Edition 2005
Reprinted 2009

ISBN 0 7282 0482 7

Typeset in Palatino 10/12 by Amy Boyle
Printed and bound in Great Britain by Bell & Bain Ltd., Glasgow

Contents

Part 2 Lease Management

Part 3 Positive Management

Foreword

As a lecturer at Bristol Polytechnic/University of the West of England 25 years ago the opportunity arose to write the first edition to meet a need created by the development of Estate Management degrees. This initiative was encouraged by the Estates Gazette.

The second and third editions were written with increasing help from academic and professional colleagues from King Sturge and the experience gained as to how to provide a practical text.

The Estates Gazette recently confirmed that demand existed for a fourth edition. This has only been possible to achieve by the College of Estate Management, under the supervision of Anthony Banfield, taking over responsibility for the book. The balance between breadth and depth remains the challenge we have both wrestled with over the years.

I have been pleased to work with Anthony and contribute two chapters. It is particularly appropriate that CEM now hold the copyright, some 40 years after my starting their undergraduate course which gave me and thousands of others, a career.

Tim Stapleton
June 2005

Preface to the Fourth Edition

The previous edition (known as *Estate Management Practice*) had become rather out dated as a result of changes in practice, legislation and government policy. In answer to my plea for a new edition, the original author, Tim Stapleton, very generously agreed to donate his copyright to the College of Estate Management if I would edit the new edition, with his assistance. On behalf of the college and myself, I should like to take this opportunity to thank him for his kind and generous gesture.

In order to reflect current market practice and the growing importance of corporate real estate occupation, I have taken the editorial decision to rename this edition of the book — *Stapleton's Real Estate Management Practice*.

In editing this new edition of the book I have generally kept to Tim Stapleton's original format with Part 1 remaining largely an historical review. However, by agreement with Tim, I have removed the Residential Property Management and Tax Planning chapters; my reasoning being that these have now developed into specialist free-standing areas and that the legislation in each area has now become so complex as to be best left to books solely covering these fields. The removal of these chapters is absorbed with the need to cover the plethora of additional legislation, much of which is covered in Chapter 6, since the previous editions. I have also added an appendix summarising some of the strategic issues for management surveyors, which I hope will be useful and stimulate thought on the strategies available when dealing with various management matters, and given greater emphasis to matters of corporate real estate, throughout the text.

As in previous editions, the updating of the book has only been made possible by the contributions of a panel of contributors and the assistance, patience and advice of colleagues, friends and family. I must give special thanks to Tim Stapleton for his help, advice and contributions to preparing this new edition.

The specific contributors are set out below:

Chapter 3 Rewritten by Bob Thompson of RETRI and College of Estate Management.

Chapter 4 Updated with additions by James Smith, semi-retired from Civil Service College.

Chapter 5 Updated by Tim Stapleton.

Chapter 7 Updated by James Smith.

Chapter 8 Rewritten by Tim Stapleton and his colleagues in the Bristol Office of King Sturge LLP.

Chapter 10 Rewritten by Steve Tyler (Nottingham Trent University).

The timing of this edition has enabled the inclusion of the changes of the Landlord and Tenant Act Part II in 2004 and other material up to May 2005.

Anthony Banfield
June 2005

Table of Cases

Table of Statutes and Statutory Instruments

Part 1

Urban Estates

1 Growth of Urban Estates in the UK

Introduction

Land ownership has, over many centuries, brought opportunities to influence a wide range of economic, social and political aspects not directly related to the land itself. Some estates such as those of the church, the Crown and livery companies have existed since the Middle Ages with only limited changes in their land-holdings. Others have been created in a generation, due to the energy of one individual and special factors unlikely to be repeated in the foreseeable future. Urban estates have been shaped by the interaction of a number of relatively well-recorded historical pressures, though different interpretations can be placed upon the motives of the parties and the merits of the outcome at various decisive stages. Economic and social development, often represented through changes in case law and statute, have adjusted the balance of rights and duties between occupiers, owners and the general public, with serious financial consequences for the groups involved. Estate managers acting on behalf of each of the parties need to interpret the factors influencing urban property holdings in a comprehensive way, and relate these to the rationale of individual estates.

Land use and estate management decisions on individual parcels are influenced by the history of occupation in both a physical and legal sense, each generation reacting to new opportunities and simultaneously creating new restrictions. The Ancient Monuments and Archaeological Areas Act 1979, is a pertinent reminder of the obligations attached by the nation to the recording of the past as part of the redevelopment process. The Environmental Protection Act 1990, Environment Act 1995 and Disability Discrimination Act 1995 illustrate newer restrictions.

Estate management practice is influenced by procedures developed over many years and exercised by a profession evolved over several hundred years with all the strengths and weaknesses inherent in such groups. The success of the landlord and tenant system has resulted in a smaller proportion of owner-occupied commercial property than elsewhere in the Western world and as a consequence highly developed procedures for the representation of the interests of both landlord and tenant and the implementation of their agreements.

The many subtle factors influencing the growth of the profession and the complex historical forces shaping urban estates can only justify limited space within the context of a study of estate management practice. A convenient, but not necessarily comprehensive, approach has been used whereby up to 1945 this has been viewed more in terms of the growth of the profession and thereafter more in terms of the effects of the factors shaping urban estate management practice.

Growth of urban estate management

In mediaeval times, the great families, the church and the Crown placed considerable responsibility and sufficient trust as they considered prudent in their stewards, who had overall charge of the management of their estates, with bailiffs physically directing and controlling the management of individual parcels of land.

Some advice issued to stewards in the 13th century suggests that the principles are unchanging:

> The seneschal of lands ought to be prudent and faithful and profitable and he ought to know the law of the realm, to protect his lord's business and to instruct and give assurance to the bailiffs who are beneath him in their difficulties. He ought two or three times a year to make his rounds and visit the manors of his stewardship, and then he ought to inquire about the rents, services and customs, hidden or withdrawn, and about franchises of courts, lands ... and other things which belong to the manor and are done away with without warrant by whom and how; and if he be able let him amend these things in the right way without doing wrong to any, and if he be not, let his show it to his lord, that he may deal with it if he wish to maintain his right.

During the 16th century changes in social and trading conditions and the rise of a growing number of smaller landlords created an environment in which the science of the measurement, recording and presentation of boundaries was able to develop. A book of 1582 entitled *A Discovery of Sundrie Errours and Faults Daily Committed by Land Meters* indicates that even then some criticism was being made of the work of surveyors. No doubt the author, Edward Worsop, was considered a very dangerous fellow; he even suggested training, examinations and the need for a license to practice!

By the close of the 16th century the rapid rise in the demand for land measurers resulted in the surveyor acquiring a reputation as an inquisitive landlord's man, much distrusted by commentators of the time. It was a logical progression from the mechanics of the making of maps and plans to their use for the purposes of the management of estates. This happened relatively slowly and it was not until the middle of the 18th century that business came forward in sufficient quantities and in a form which enabled practices to be created and sustained beyond the lifetime of individuals.

Throughout the 17th century, the surveyor in his various forms was playing a relatively humble role instructed by attorneys, stewards, conveyancers and architects. During the latter part of the 18th century some lawyers devoted more and more time to their role as stewards managing estates and some land surveyors were successful in obtaining steadier income from regular management of the estates they surveyed.

The 18th century saw the start of the Enclosure movement and by 1801 a committee of the House of Commons was in favour of the appointment of a valuer as well as a surveyor in every Enclosure Bill. At the same time the Ordnance Survey was growing in importance, not without some friction between the military surveyors and civil surveyors, this state mapping service reduced the need for the private land surveyor's work on enclosures. The industrial revolution was also gaining momentum with urbanisation and the construction of, first the canals, and then the railways.

The first half of the 19th century can be identified as the period in which the surveyor's previously chief occupation, land measurement, had superimposed upon it a much wider range of activities, all of which were due to aspects of the industrial revolution. The building of the railways required individual Acts of Parliament; their promoters became skilled in the drafting of these measures and the preparation of the necessary evidence. Their surveyors had considerable impact on the individual resident agents with whom they negotiated along the proposed route. Between 1850 and 1870 over 100,000 acres were purchased for railway purposes, similar to the activity of the motorway programme of the 1960s and early 1970s. Land agents of agricultural estates found that meeting the demand for urban growth offered greater returns than agricultural rents but by applying agricultural practice where possible, retained ownership of developed land for their client, thus creating the unique London estates.

All this activity created a considerable demand for the skilled surveyor, and while the eminent grew in reputation many incompetent persons sought to exercise the surveyor's skill. In the same year that the now RIBA was created, 1834, a Land Surveyors' Club was formed in London for active promotion of the profession; they sought to exclude resident agents from membership, who were servants of a master rather independent consultants.

In the spring of 1868, 20 surveyors met in London and by November of that year had formed the Institute of Surveyors and held their first ordinary general meeting at 12 Great George Street. Three-quarters of the original members had been involved in valuation, negotiation and arbitration for railway works but all the branches of surveying were represented within their practices. Some of their practices in both London and the provinces had been established for a century before the formation of the Institution and its member firms possessed considerable maturity.

The most eminent of the 19th century surveyors was John Clutton whose life spanned almost the whole century. Between 1845 and 1851 he was appointed surveyor for the southern half of England for the Ecclesiastical Commissioners, Adviser on the Royal Forests and Crown Receiver for the Midlands and the South of England. A close second was Robert Collier Driver, who between 1860 and 1875 was responsible for a million pounds of property sales per annum. In Bristol, William Sturge was equally active, acquiring land for the Bristol and Exeter Railway, Somerset and Weymouth Railway and South Wales extension of the Great Western Railway.

The latter half of the 19th century saw tremendous growth in urbanisation as the new municipal authorities acquired land for local urban infrastructure in much the same way as railways acquired land for the national infrastructure. The Surveyors' Institution had considerable influence on policy, legislation and administration and was mentioned in the Metropolis Management and Building Act (Amendment) of 1876. The surveyors gained a charter in 1881, which assisted them in their aims of intellectual advancement, social elevation and moral improvement.

The Institution was represented at the Congrès International des Géomètres held in Paris in 1878, where it was suggested that, in order to practise, surveyors should be required to obtain a government diploma; the rest of Europe accepted this principle. This did not accord with British practice, but the surveyors did set up a system of qualifying exams for new members in time to support the application for the charter.

Codes of conduct and the role of local branches were developing in the early part of the 20th century. The Liberal Budget of 1909 posed questions on land taxation which really only had to be faced 40 years later; though by then professional bodies had appreciated the need to avoid arguing their case on political grounds.

During the 1914–18 war agriculture and industry flourished in an effort to meet the needs of war. Speculative development of owner-occupied housing, reflected the concepts of the garden city movement of the 1920s, and the start of the decline of the rented housing stock. At the same time among the then two dozen or so publicly quoted property companies the seeds of commercial and industrial development were being sown, the latter often based on War Department surplus. Despite 3 million unemployed in the early 1930s large office buildings were constructed in London, the subject of major refurbishment in the 1980s. Industrial estates were developed in major urban centres, many of which have since been redeveloped.

With hindsight the 1920s and 1930s can be seen as a time when professional bodies enjoyed a period of relative stability in which to develop, particularly at branch level. The ownership of land was not a major issue and there was fragmentation of both urban and rural estates which offered new opportunities for development carried out mainly by builders utilising traditional sources of finance. Towards the end of this period a range of social pressures in health, planning, education and welfare were building up, but their resolution was delayed until after 1945.

Education

At the end of the 19th century some of the more fortunate articled pupils were able to take advantage of the educational facilities which had arisen as an extension of the London practice of Parry, Blake & Parry. After the first war and some inter-institutional rivalry between the Chartered Auctioneers and the RICS, which did not reflect well on the latter, the College of Estate Management was formed in 1919 and absorbed the educational facilities of the above firm. It received its Royal Charter in 1923.

The early date at which the college was able to teach for an external degree of London University may not have been entirely unrelated to the fact that the Vice-Chancellor was the brother of Sir William Wells, a past RICS President. The Cambridge degree, then predominantly agricultural, was being developed at the same time. By 1947, the college had a sizable full-time student body in addition to its postal courses for the exams of professional bodies. On its move to Reading in 1972, its full-time courses became an integral part of Reading university while the college kept the postal correspondence course students. The Watson Committee of 1950 recommended that the RICS should encourage the provision of full-time courses, and technical colleges in strategic towns were encouraged to provide courses. The creation of polytechnics in 1969 incorporating the major technical colleges which had been teaching for the RICS exams since the 1950s enabled further growth to take place in the provision of full-time courses. By 1977 the majority of those entering the profession had qualified through full-time courses, fulfilling the prophecy of the Eve report of 10 years before. In the 1990s the RICS introduced an accreditation scheme, subsequently to become a partnership with universities and colleges throughout the world able to run accredited courses and has applied minimum entry levels for students to enter the courses.

This has enabled the profession to attract some of the most able of school-leavers necessary to take it forward with confidence in its ability to handle the complex issues of the next decades. Post-graduate courses (often at Masters level) now enable semi-cognate and non-cognate first-degree holders to enter the profession, and the need and the means for continuing professional development (also known as Life Long Learning) has been achieved. The change in balance between cognate first degree holders and post graduate entry into the profession has been demonstrated by the RICS statistics on student membership.

The development of education for the profession has been more rapid than any other comparable discipline, and this has not been without its controversy. The perception of our Victorian forebears in anticipation of the debate is illustrated by a brief extract from the views of William Sturge in 1868, in warning against a university education for young surveyors, "... the tastes and habits he will form will probably render the drudgery of a surveyor's office distasteful to him. I arrive at the conclusion that the balance is against a university education for the surveyor". This can be contrasted with the view of Jeremiah Mathews, who, two years later, was led to the conclusion that, "... a university education would avoid a too premature technical training ... which warps the mind by confining it to a single channel of thought and thus renders a man incapable of understanding the motives which guide others".

Post-1945

In the post-war period fundamental changes have occurred across the whole fabric of society. Five changes have particularly influenced the way we live and, as a result, the demand for buildings and the way they are used:

- educational and training opportunities for young people are much greater and the young have greater mobility and purchasing power
- many more women now go out to work, with most families having more than one income and a consequential dramatic fall in the birth rate
- a greater emphasis on the creative use of leisure time with many specialist community-based facilities
- the adoption of ideas from other countries and other cultures as a result of increased travel opportunities, reporting by the media and immigration
- the impact of technology on the way we live, work, shop and use our leisure time.

This happened within a system of government and a planning framework with a much higher regard for social priorities and a growth of public awareness and knowledge of environmental issues. Time-consuming consultative procedures developed which in periods of high inflation rendered capital budgets of projects almost ineffective. These features have been particularly marked in the provision, funding and use of our housing stock. In the immediate post-war period and particularly the early 1950s, the quantity of housing (ie not the quality) was regarded as something of a triumph. However, the problems of the tower blocks built in the 1960s caused by the application of technology running ahead of the policy-makers understanding of human responses has subsequently resulted in a more introspective view. Since the middle of the 1960s we have seen a continuous fall in housing starts, possibly due to the political parties using housing finance as a political battlefield.

The changes in housing over the last 50 or so years have been startling, whether measured in terms of consumers or of the nature of the fabric. Between 1951 and 1978 the proportion of one-person households doubled to 22% of the total number of households. Over the same period, the number of households without a fixed bath or sharing a bath fell from almost 50% to 4%.

The UK has had one of the greatest variations in the distribution of wealth in Western Europe. This is despite during the 1970s the severest redistributive income tax, which has had some effect; for example, the proportion of wealth held by the most wealthy 1% of the population fell from 33% in 1966 to 24% in 1977. However, in 1977 the wealthiest 10% of the population still owned over 60% of the wealth, and surprisingly the poorest 50% of the population owned only 5% of the total wealth. The government's statisticians preparing "Social Trends" do not claim infallibility in absolute terms, and the issues are complicated by the patterns of social welfare and the black economy, nevertheless the overall trends are not in dispute and have long-term implications for estate management.

The taxation of unearned increment is mainly justified on social rather than revenue-earning grounds. Development Land Tax only provided a modest revenue but such taxes can have an effect on land management out of all proportion to their revenue-earning potential. The stones dropped in the pond by the legislation of 1947, 1967, 1973 and 1975/76 sent ripples into every aspect of estate management, directly influencing the development process and hence the nature of the flow of property for occupation and investment. Conflicts between demand for occupation by users and demand for investment purposes by institutions have directly contributed to the continuous cycle of boom and slump in construction and investment which prevailed up to the late 1990s.

Since the middle 1950s the institutions, initially insurance companies and subsequently pension funds, have taken up direct investment in property. At the, end of 1977 the aggregate property holdings of insurance companies were £7,153 million, almost double that of the pension funds, though pension funds as a whole and their property portfolios in particular were growing much faster than those of the insurance companies. By 2002, financial corporations in general owned £83 billion of commercial property, although, as will be seen in Chapter 3, the actual percentage that property ownership represents within insurance companies' portfolios has decreased significantly. In the past, over the long term, property has offered significantly better returns than gilts or equities, measured in terms of income and capital growth, though there are some worrying aspects. The market has traditionally been dominated by the sentiment of no more than a dozen major funds and property is really only better than other investments if you do realize the benefit of the capital growth. What do you then buy with the proceeds? Some very large portfolios have been created with a minimum of disclosure to interested parties, containing high-value properties due to new building techniques and materials permitting the construction of large floor areas to meet user requirements. On occasions the techniques and materials showed some unforeseen and expensive maintenance problems.

The calls for new vehicles for investment in property in the UK with tax advantageous schemes led to government consultation on plans to introduce a Property Investment Fund (PIF) structure which was similar to that of the Real Estate Investment Trust (REIT) structure introduced in the US as long ago as 1960 and which has increasingly been adopted (with local variations) in many countries throughout the world. In the 2005 Budget, the UK government confirmed its intention to introduce such a scheme (now to be known as REIT) in 2006. Because of the UK's history of public real estate companies REITS are likely to be a more natural vehicle for investment in the UK than they have been in the US where they have mainly concentrated on larger sized developments. The fact that US REIT capitalisation grew from under $10 billion in 1993 to over $200 billion by 2003 and the adoption of such schemes in countries from Germany to Singapore suggests that such a scheme should prove popular to UK investors, if set up properly, and should attract more investors.

Historians with more perspective than we enjoy may well not be impressed by any of the above as the key factor affecting estate management in the last 50 or so years. The oil crisis of 1973 (and the 1991 Gulf War and oil crises) caused a destabilizing effect on the whole range of economic activity, bringing in its wake record interest rates, rapid inflation, particularly in construction costs, and periods during which there were falls in the standard of living of the average tax-payer. This resulted in the destructive degearing of industry in general and property owning organisations in particular. Within property management the increasing cost of energy has raised service charges from an incidental cost to a major component of the costs of occupation. Designers are now seeking to respond with low-energy technology and the new discipline of terotechnology has emerged.

Sustainability has been accepted by successive governments since the Earth Summit of 1992.The concept and principles are contained in the UK government publication *A better quality of life: a strategy for sustainable development of the United Kingdom*. The strategy has four main aims:

(1) social progress which recognises the needs of everyone
(2) effective protection of the environment
(3) prudent use of natural resources; and
(4) maintenance of high and stable levels of economic growth and employment.

Sustainability has become a new mantra for planners, occupiers, politicians and designers alike. The apparent intellectual idealism can be expressed in terms of six criteria:

- longevity
- loose fit
- location
- low energy
- likeability
- loveability.

These will be interpreted by stakeholders; those who take decisions and those affected by the decisions. The action by the former and consequences for the latter are influenced by the extent that policy or even better legislation can "internalise the externality" and so bring the interests of the two groups together.

Stakeholders will tend to approach the issues from one of three disciplines.

- *Economic* — direct internal financial consequences of procurement or occupation or their broader external economic impact.
- *Environmental* — this is principally about conservation of natural resources, hence the efficient use of energy, this can be measured by calculating the embodied energy in processed materials to create buildings and so emphasises the energy saving from refurbishment. External factors are illustrated by water, waste and transport costs
- *Social* — recognised over the last few years by Corporate Social Responsibility.

In order to apply this triple bottom line, Sayce, Walker, McIntosh (*Building Sustainability in the Balance*) have developed a Building Sustainability Assessment Tool which scores the three disciplines from the point of view of Internal and External Stakeholders. Perhaps only government as a major owner-occupier and uniquely representing all external stakeholders can set the standards.

Of all the factors influencing investment, conventional wisdom may well be the most important. A Department of the Environment report (1976) with the inelegant title *The Recent Course of Land Property Prices and the Factors Underlying It* sought to analyse the many possible factors, but concluded in its final paragraph:

> The major explanation of the property price boom which we are left with, but which is regrettably difficult to test, is that expectations about the future played a vital determining role.

Putting it simply, a small number of influential opinion-formers can have a quite disproportionate effect on the market, the nature of which is a vital input into the decision-making process in estate management.

The Conservative Government elected in 1979, had, with its fourth term in 1992, achieved its objective of a fundamental shift in capital resources from the public to the private sector. The subsequent Labour Governments continued this trend. The privatisation of utilities and the introduction of private sector techniques into the management of public sector estates coined a new phrase "property asset management", but the principles remain those of the 13th century land steward.

The recession, which commenced in 1989 and failed to respond to the advocacy of recovery, certainly ran for four years, the longest since the 1930s. This has meant the experience of the 1980s based on steadily rising property values, which was fully used to underwrite loans became a handicap in the 1990s. The result was that the most fashionable became the most vulnerable, the least fashionable, the most secure. Banks saw their debt converted into often worthless equity over a period of 18 months.

Since 1996, inflation has been kept down, partly as a result of the setting of the minimum lending rate being depoliticised by being handed over to the Governor of the Bank of England to review and set.

Commercial and industrial property vacancy rates, which peaked at 11% in 1991/92 had fallen to 7% by 2001/02 (source ODPM website *www.odpm.gov.uk*).

Individual estates

The ownership of property may be sought for many reasons. Effective estate management requires a clear knowledge and understanding of the motives of the client, whether that is the estate owners or the occupiers. Estates are held or occupied in order to fulfil a need; if that need is clearly enunciated then the estate manager can use his skill to manage, refurbish and develop or dispose of the estate in the most desirable manner.

Interests in property are a means of storing wealth; this may be as capital asset producing no income, such as a development site, or as income from a fully developed site with an opportunity for real growth. In general, trading and commercial companies' property is a fixed asset and the best security against which the company can raise loans in order to further its business activities. The effect of estate management policy on the value of the assets, and hence the security offered to the investor has, in periods of unstable values, resulted in guidance from professional bodies on the procedures to be adopted to give a true and fair picture of this store of wealth. Such procedures and regulations affecting property emanate from national and international bodies such as RICS, Institute of Bankers, International Financial Reporting Standards (IFRS) and International Accounting Standards (IAS).

Newly formed businesses usually rent property, but there will be uncertainty as to its continued cost and occupation. Ownership offers independence but brings with it additional responsibilities and the possibility of inefficient use, due to the lack of financial discipline imposed by reviews of the cost of occupation through rent-review clauses.

But ownership of property, if that is not the core business can tie up capital unnecessarily and many large retailers such as Currys, Tesco and even Marks and Spencer have sold off some of their freeholds and taken a lease back.

The ownership of property is no longer a necessary condition for the election of representatives to government. Nevertheless, it certainly remains in the eyes of many a prestigious acquisition and may bring with it social and personal advantages as well as an element of kudos. Individuals can become very attached to their physical environment and the continued association of the family or business with particular land and buildings provides some certainty in a world of change, though the high maintenance costs, taxation and custodial tasks which come with heritage property impose severe burdens on owners.

However, the conduct of some estates and property interests has acquired a certain notoriety, with some landlords falling below the standards imposed by public opinion or the law.

The widening responsibilities undertaken by the state since 1945, subject to varying emphasis by different governments, require specialist properties, managed to meet the operational requirements of public sector bodies. The public sector is the owner of about 20% of the surface area of the UK, though admittedly nearly half of this is the operational land of the Forestry Commission. Apart from its specific land-ownership, the state also possesses considerable powers to influence the pattern of land-use, the management of urban property and the obligations associated with the ownership of land and buildings. This concept of national estate management needs to be distinguished from the state as a group of large land-owners.

The national estate

The national estate, and by implication the need for its management in national terms, is one which surveyors have tended to ignore as anathema to their essentially historical private sector philosophy. It is often difficult to discover the pattern of land ownership, and town planning has tended to concentrate on what can be seen or measured — the control of land use — to the exclusion of the proprietary land unit. The concept of the management of the national estate is therefore a much greater one than the statutory planning system concerning itself primarily with controlling change of use within a county structure plan and district local plans.

Bearing in mind the position of the Crown in the 11th century as the ultimate landlord and the role of land as a tax base, then for a country under new management, a national estate terrier, the Domesday Inquest of 1086, was an essential and obvious measure. The only other comparable survey of proprietary interests was based on the rating returns of 1872/73, when agricultural land was still subject to rating. This listed the name and address of owners of an acre or more of land; such a basis had little relevance in urban areas. The survey showed that 57% of England and 93% of Scotland was held in estates of 1,000 acres or more. Since that date until the late 1950s there was an unprecedented period of fragmentation of estates, then the financial institutions began to add direct property holdings to their portfolios. In conceptual terms their role is identical to that of the old-style charities, acting as nominees to ensure our rights to a reasonable income in old age to supplement what the state may provide. From originally being just another investor, they have become the single most important factor underlying demand in the property investment market.

But what can we discover of the nature of the national estate? The agricultural sector is probably the best documented by physical area, tenure and quality, with a very small annual percentage change in ownership. The Northfield Committee of 1979, despite making many detailed recommendations, found no fundamental defects in the existing pattern of ownership.

The diversity and size of the residential sector, representing the vast majority of the nation's store of wealth, makes it the most difficult to handle. This was made much more difficult by the advice of Secretary of State for the Environment in the summer of 1979, to 19 million ratepayers to tear up their rate return forms. However, there are some statistical sampling surveys of motivation of potential occupiers and quality of the stock. Local authority/housing association and voluntary sector housing tends to be relatively well documented.

As regards commercial and industrial property, the total cost of net acquisitions by institutions is well documented in "Financial Statistics". The ODPM (Office of The Deputy Prime Minister) and its predecessor, DOE, have collected various figures, mainly based on aggregation of the records of individual offices of the District Valuer/Valuation Officer, as has the British Property Federation (BPF). According to the latter, the total value of all commercial property in the UK in 2002 was estimated at £3.31 trillion, including residential; or £565 billion for just the stock of commercial property. In that year commercial property represented 5.9% of GDP. To the extent that exceptional progress has been made in the field of asset valuation, the aggregate of values ascribed to properties in the private sector in company annual reports may now offer a more up-to-date statistical source.

There is no common basis underlying any of the statistics in respect of agricultural, residential or commercial property, and to these must be added a complex and specialist public sector estate. Parliamentary pressure on government resulted in the first annual report of the Property Services Agency in 1978. The estates of local authorities and statutory bodies had been relatively unknown outside the committee room until the 1980 Local Government Planning and Land Act introduced registers of surplus public sector land, which coincided with government pressure on public bodies to review all their operational estate. This was followed by the setting up of Property Advisers to the

Civil Estate (PACE), succeeded by Office of Government Commerce (OGC), to co-ordinate government land ownerships.

The question which must be considered is whether government policy and implementation in respect of matters affecting land are simple individual piecemeal expedients or whether there is a measure of overall policy and an availability of information against which it may be possible to assess that policy? At this stage political issues may be raised, but not in conflict; there is a remarkable unanimity of support for the need for such an approach. Both Dr Denman, of Cambridge University the eminence grise of the right in matters affecting land, and the Centre for Environmental Studies, have identified the importance of proprietary interests and the lack of information.

Even in the 21st century there is still a lack of any one good database for identifying trends in the pattern of land-ownership, assisting policy decisions on land-use, determining the availability of land for development, testing the achievements of the planning system. However, the Land Registration Act 2002 and Land Registration Rules 2003, which contained provisions for electronic conveyancing, should in time make the transfer of land easier and enable such statistics to be more readily available.

In view of the problems facing the private estate owner in assembling the information necessary to be able to take estate decisions with confidence, so the problem facing the State in deciding upon the correct response to perceived problems must be immeasurably more difficult.

The First Land Utilisation Survey in the 1930s, by the late Dr Dudley Stamp, was one of the factors leading to the introduction of the comprehensive post-war planning system. Following Capital & Counties sponsorship of a feasibility study into the creation of a National Land Information Service in 1991, the service is available on the web but even in 2005 it was still not fully developed. It is part of the government's strategy for "electronic service delivery" under the Central/Local Government Information Age Concordat. The ultimate aim is to access from a central hub, data held by:

- government departments
- remotely sensed data
- local authorities
- private sector databases
- socio-economic databases
- utilities
- land registries

- Inland Revenue
- Ordnance Survey.

The research departments of the major private practices have since the late 1970s made a significant contribution in publishing market intelligence in respect of:

- micro-economic studies
- investment performance measurement
- stocks and the utilisation of commercial property
- user requirements.

The IPD (Investment Property Databank) was set up through sponsorship by a number of the main commercial surveying practices in the UK to provide impartial information such as analysis surveys and, in particular, indices from rental levels and capital values of commercial properties. The service now covers other European countries. It is a wholly independent company with the major equity in its ownership retained by its founders. By 2003, the 10,811 commercial properties covered by the IPD Annual Index in the UK totalled £105 billion.

In addition, greater awareness and confidence between academic institutions and practice, supported by the Research Councils has helped to initiate joint research projects. These offer the advantage of bringing together the best of the market and client motivation of practices, with the methodology, rigour and longer term analysis of academic research. Estate management has been the poorer in the past due to the inadequate research base compared with other vocational areas of study.

The most rapid rate of growth of any land use in the post-war period has been dereliction. The combination of the surveyor's very success in creating the best financial situation for his client, together with a planning system obsessed with the statutory plan-making process, plus a rapid rate of technological change, laid waste to some of our major conurbations on the scale of war-time bombing.

Thesis and definitions

The basis upon which this book is founded is that the essentially technical process of property management has superimposed upon it two further tiers of activity. First, estate (or asset) management, and this means identifying the role of property within the broader overall aims of the particular entity of which the land-holding forms part. In many cases property will be simply one of a number of resources of an organisation, and this may require a more flexible application of estate management skills than those appropriate" in a solely property-orientated management. Second, a level of public involvement in the private sector through various agencies, statutes and interest groups sufficient to justify the concept of national estate management. Linking the two, a growing sense of financial accountability in the public sector and sustainability, which raises some doubts as to the validity of traditional differentiations of purpose and objective between public and private sector estate management.

It is always difficult to face up to the definitions. As a first step it is helpful to distinguish between the following definitions.

Estate management (the generic activity) is defined in an RICS Policy Review in 1974 as:

> All facets of the use, development and management of urban land, including the sale, purchase and letting of residential, commercial and industrial property and the management of urban estates; and advice to clients on planning.

Estate management (the specialist activity) is defined by Thorncroft as:

> The direction and supervision of "an interest" in landed property with the aim of securing the optimum return; this return need not always be financial but may be in terms of social benefit, status, prestige, political power or some other goal or group of goals.

This is a very helpful definition and the first two chapters of this book are designed to enable the reader to interpret this as fully as possible, inserting his own value judgments and perceptions of estates and clients. Arnison has criticised Thorncroft's definition as springing from a totally private sector client framework and would substitute the phrase "the national estate" for "an interest" in the above definition. He goes on to identify three quite different types of personnel with a role to play in estate management. Skilled practical men (and women), accounting for the majority of those in practice, providing a service in the public or private sector, in valuing, managing or disposing of everyday property interests and property problems. A smaller, perhaps more mentally agile, group managing more complex situations, some distance removed from individual property interests, including senior partners and directors of the larger London and provincial partnerships, senior staff in the public sector and the directors and executives of the major property and investment companies and institutions. Lastly, a small group providing the innovation, the critical appraisal and reappraisal of methods, techniques and attitudes which ensure the vitality and adaptability of a profession. Perhaps there is a little of each in all of us?

Sources and further reading

Building Sustainability in the Balance, Sayce, Walker, McIntosh, Estates Gazette 2004.

Chartered Surveyors — The Growth of the Profession, Thompson, Routledge and Kegan Paul, 1968.

Commercial Property Key Facts 2003, Andrew Scott London Business School for British Property Federation (see *www.bpf.propertymall.com*)

Social Trends, HMSO.

The Place of Commercial Property in the UK Economy, The Farmland Market, Estates Gazette/Farmers' Weekly. London Business School, 1991.

Understanding the Property Cycle, RICS, 1994.

Whose Land is it Anyway, Norton Taylor, Turnstone, 1982.

Management 2

Introduction

Management of any type, whether of property or of any other resource or process, is concerned with what happens between the main decision-making structure, whether it be the board of a company, trustees, an appointed commission or an elected authority, and the performance of the operational task. In the case of a small business or partnership, since there will be little delegation, there need be little formal control other than that achieved by the normal flow of correspondence. One of the main functions of management is the control of devolved responsibility. Many small organisations without such devolved responsibility achieve their success or failure solely on the personal qualities and skills of their proprietors. This is not to say that such organisations will not find the tools of management useful, rather that a detailed functional approach to management may not be very relevant. To the extent that a larger proportion of property is now being managed within large organisations, the case for a study of management in its own right becomes stronger. Management has been defined by Allsopp as:

> The selection of goals and the planning, procurement, organisation, co-ordination and control of the necessary resources for their achievement. Management is concerned with the dynamics of circumstance and activity and is generally motivated by the need to economise in the use of resources and time in achieving predetermined objectives. In commerce, management is concerned with efficiency in the conversion of opportunity and resources into wealth.

The operational heart of this can be expressed as converting policy into action, leading to the questions:

- What policy?
- What action and by whom?
- Is it the right action?
- What are the long term effects and implications?

Much of the remainder of this book is concerned with applying these questions to estates and individual properties. This is achieved in Part I by an analysis of the different kinds of estates, in Part II by decision-making in the management of leases and in Part III by decision-making in relation to estates. There are many interpretations of management and its theories have developed within business management as an essentially product-orientated approach, with management having somewhat combative implications within the industrial relations field. In order that a study of management makes a positive contribution to estate management, considerable care needs to be taken not to go too far down the business management route. Business management is becoming more a social process with an emphasis on people management within large organisations producing a standard product in very large quantities, indeed, a growing capacity to over-produce in many technological fields, whereas estate management is concerned primarily with the stock of a heterogeneous specialised asset; people are important, but the numbers in all but a handful of organisations are relatively small, and the task is one where standard procedures can only be applied with the greatest of care, education and training. In fact many of the problems that arise require a combination of perceptive originality and experience and at any one time only a limited number of individuals will be in a position to exercise such qualities. Quite apart from the diversity of the types of property and estates, the organisations which carry out the management function have very different characteristics in terms of rationale, size, mode of operation and accountability.

To summarise, the problem would seem to be agreeing upon some general functions of management and relating these to property, bearing in mind that the education of the majority of surveyors up to the mid-1980s had tended to concentrate on skills of how best to carry out specific instructions rather than contributing to the writing of instructions, although this has subsequently been addressed. Also, other than in specialist property organisations, surveyors have at times been reticent in actively integrating the estate within the overall management function and structure of the parent organisation.

On the consensus of a number of authorities considered by Thorncroft it is proposed to examine the general functions as planning, organising, co-coordinating and controlling.

The application of these general functions of management to property is not without its difficulties or its critics. This is partly due to considerable misunderstandings and also a fairly standard human failing among professionals of a dislike of their expert opinion being challenged, particularly at an interface with another discipline. If the general principles of management can be broadly stated then this enables the individual surveyor to apply them to any aspect of estate management, utilising his own experience to the full. Appendix A contains a flow-chart illustrating some broad aspects of management within a large corporate organisation.

Planning

This relates to the work of the policy-making group within the organisation, concerned with forecasting on different timescales and the analysis of all the underlying factors affecting the organisation activities or potential activities and determining any necessary responses. A few, or even one leading personality may create policy, alternatively there may be a corporate approach. In whatever way it is produced or structured or communicated, the planning function having determined policy should then be expressed as a management brief, up-dated with the benefit of further information that arises from regular monitoring of performance. This is considered in Part III, which has as its theme positive management, and particularly by Chapters 7 and 10.

One recent problem is the need to plan further and further ahead, due to the lengthening gestation period of projects of every type, but the factors underlying the decision seems to be changing more quickly and more significantly than past experience is readily able to assimilate. This then leads not necessarily to a counsel of despair but to opportunist planning, which can be efficient in its individual execution but less effective in its overall impact. This is likely to result in the planning process becoming more important than the plan itself; that is to say the process of planning is justified as the best way of ensuring that all factors are considered and agreements made as to the weight to put upon them and as a consequence their effect on decisions of the policy-making body.

Organising and co-ordinating

This is concerned with setting up a structure to perform the management brief, and there are several approaches. It may be a case of the executive manager receiving a brief and using his experience and knowledge of his staff and resources, deploying them to meet that brief. Alternatively, where an opportunity exists for change or some new venture, a number of issues can be identified:

- specialist or generalist organisation?
- centralised or decentralised organisation?
- effectiveness or efficiency?

The last of these is particularly interesting. Put simply, efficiency is concerned with ensuring that each particular operation is carried out to yield the most desirable cost/value ratio. Whereas effectiveness implies standing back from the operation, ensuring the objective is clear and then determining which of several alternative methods is most able to meet that objective. Discussions of this type tend to result in rather heated meetings, a sure sign that personalities and issues have become interlinked. The above usually has to be carried out within an existing organisational structure with its own management style, threshold of responsiveness to change, pattern of resources and staffing.

This leads into the third function of management, the coordinating of staff and resources into the most effective means of operation, integrating effort. This is closely related to the selection and motivation of staff and that almost indefinable atmosphere which does exist in efficient, effective and well-managed organisations. A study of the way that policy once formulated is disseminated throughout the organisation, and the way that the right information reaches policy-makers can go some way to establishing the importance of the co-ordinating function. Specifically, this depends upon communication, both the means of communication and the effectiveness with which the information is handled and directed to those most able to use it.

Control

Control is only necessary if there has been delegation and this leads to the conclusion that the smaller the unit of accountability or profit centre the less relevant is a structured pattern of management.

Assuming a sufficiently large organisation for the four functions of management to be distinct, then effective control is more than the receipt of reports, however detailed and accurate they may be. It is the measurement of the performance indicated by the information in the reports against planned or desired objectives. The means that are necessary to achieve effective control will vary significant between different management situations. What is essential is that monitoring methods, give appropriate feedback to the policy-makers as they enter the next stage of their planning with more information and so up-date the brief.

At its simplest and possibly worst, the brief may be no more than a company's annual report with an analysis of the past year and a prospective of the future. It could be much more, not just a management tool but part of the system of motivation that ensures that a member of the organisation never needs to ask, "what should I be doing?" nor wonder whether he has done it right, for at the end of the day, the only real management tool we have is those who work around us.

The relationship of management to property

The most useful contribution of general management to property is the theory of "Management by Objectives", developed by management "gurus" in the early 1960s. The case for such a theory has been stated by Druger as:

> what the business enterprise needs is a principle of management that will give full scope to individual strength and responsibility and at the same time, common direction of vision and effort, establish teamwork arid harmonise the goals of the individual with the common weal. The only principle that can do this is management by objectives and self control.

The process consists of:

- management formulates strategic plan by defining corporate aims and objectives in the short, medium and long term
- action and resources are co-ordinated within a tactical plan. Individual units then prepare means of implementation, with realistic and measurable performance requirements for individuals
- regular monitoring and review is undertaken on a appropriate frequency. This is set out in diagrammatic form in Appendix A,

and is readily applicable to any large estate management organisation. It is also relevant to the refurbishment cycle of any large commercial building, such as a shopping centre or major multi-let office building.

Four distinctive types of estate management can be identified:

1. The management of the proprietary unit exercised by: directors, elected members, trustees, partners, public servants, receivers or others in whom ownership is vested. In any organisation in which property is not the primary purpose, the main skills will be those of corporate management with an emphasis on business and financial management.
2. The management of the estate management organisation, requiring the application of business and personnel management skills within a particular professional setting. In the case of property investment companies and specialist professional practices, this second type of management will be almost indistinguishable from management of the corporate organisation, and is particularly covered in this chapter.
3. The management of specific:
   - estates/portfolios — with operational or investment objectives
   - areas — geographical areas
   - functions — types of professional activities.
4. The management of individual properties or leases, and Part II is concerned exclusively with contractual and statutory aspects of the occupation of leasehold property, with Part III concentrating on a wider range of activities.

The great majority of the training and practice of estate managers is concerned with the last of these four, though lack of understanding of the other three may relegate the practitioner to the status of a technician. Indeed the RICS published a report in 1985, entitled *Competition and the Chartered Surveyor* which identified practice management as the key issue.

Considerable care has been taken to describe these four activities as types rather than levels of management — levels suggest some intrinsic merit in progression from one level to the next. This may well be the case for some individuals, but having regard to the different sizes of estates organisations and the financial importance of the correct specialist advice upon detailed aspects of property management, no career paths or education and training implications are implicit in the order in which these four types of management have been analysed.

The estate's setting

Chapter 1 traced the historical growth of urban estates and introduced examples of the various factors determining the shape and pattern of property ownership and the use and development of land. These can be classified as follows.

The social climate in which there is greater accountability for the effect individual or corporate decisions have on the community. Various practices in the commercial, and particularly the residential, fields have been judged, not by the courts, but by public opinion, and found defective. The knowledge and influence of the Fourth Estate in technical fields is strong and extensive research programmes have been carried out by various agencies into aspects of land management and land use. The effect of the social climate will vary, depending upon the identity of the estate but is effect on the motivation of staff and the response of tenants and other interested parties should not be underestimated.

Political change has characterised the post-war period and no more so than during the 1970s. The property market in its various forms as investment asset, product of the construction industry and provider of services to the population at home or work is particularly susceptible to changes in political thought. Some legislation is politically inspired in both its creation and demise as can be seen by the Community Land Act in England. Others, for example, Capital Gains Tax, blended into acceptability. More fundamental changes, such as entry into the EU and Human Rights legislation, have within them the seeds of long-term changes in our understanding of property rights.

Statutory change has put flesh on to political ideology so much so in the case of the Community Land Act in the 1970s as to cause its death through obesity. The most significant changes have not been of primarily property legislation but of secondary legislation affecting the quality of the environment and employment and business practice. This has resulted in much closer liaison between surveyors, solicitors and accountants.

The frighteningly bad economic performance of the UK in the 1960s and 1970s resulted in remedies which looked more like self-inflicted wounds. Consequently, long-term estate management plans have at times been severely distorted by high interest rates and high cost inflation, and some periods of falling or static property values. Apparently rapid rises in values need to be measured in real value terms against longer time intervals which allow for the periods of depressed values.

Physical and technical factors related to higher expectations among tenants and some serious failings of both materials and techniques have caused major maintenance problems early in the life of buildings. Inner city decay caused partly by the success of the planning policies of the 1960s and the overall scale of industrial activity has resulted in thousands of hectares of wasteland which, due to complex institutional factors, seem to many to present almost insolvable problems.

The aggregate effect of rapid change in all the above parameters of estate management have created strains on the traditional pattern of land-holding, resulting in substantial sales and acquisitions of property and the merger and demise of once flourishing organisations.

Throughout the changes in economic climate, property assets still have to be managed and the recession in the early 1990s changed the perception of the property management departments in many firms. Whereas property management had been almost regarded by some as the Cinderella department of such firms as it did not have the apparent glamour and high fee earning potential of investment development and agency departments; however, when the deals (and hence fees) dried up in those departments the property management department continued to earn regular fees. Unfortunately, this also lead to firms more actively trying to get management instructions which resulted in fees being cut (sometimes to seemingly uneconomically low levels) and in some cases the standard of management provided being reduced to basic levels with additional fees charged for all "extras".

Staffing

The estates staff will enjoy the benefit of a number of cohesive forces which will help the team to achieve the desired goals. They will be able to identify with a physical estate and the professionally qualified staff and those aspiring to that status will have a common body of knowledge and expertise. Some care is necessary in the way senior non-professional staff relate to recently qualified staff, the former often contributing a considerable depth of experience across a narrow front of the management task; the latter approaching the situation with an almost opposite mode of thought while the administrative staff tends to take its standards from the management. The balance between professional, technical and administrative staff reflects the management style of different organisations, but generally the larger the organisation, the larger the proportion of support staff. Some estates operate with an entirely white-collar staff, in others, rather like an iceberg, there is a much greater blue-collar staff performing a direct-labour role. The introduction of compulsory competitive tendering in the late 1980s has imposed a market discipline. Some direct-labour staff for continuing small maintenance tasks can usually be justified but whether dealing with direct or contract labour a high standard of supervision is necessary.

In the last five years out-sourcing has swept through the private and public sectors, with OGC (Office of Government Commerce) and OJEC (European Journal) driving the latter.

The introduction of computer systems has in many cases reduced the burden on, and numbers of, administrative staff required. However, some may argue that this has merely shifted the balance towards an increasing number of IT consultants and staff.

The portfolio

The effective management of the portfolio will require substantial contributions from other disciplines: lawyers, accountants, architects and other specialists within the construction industry. The portfolio may vary from an owner-occupier estate in industry, to an investment estate in the financial sector, to mainly leasehold occupations by a high street trading company or manufacturing company. They will all be subject to the same underlying factors, but the weight to be attached to each and the different time-scales involved, will result in a wide variety of response to changes in estates policy. A portfolio consisting of retail leasehold occupations can be changed relatively quickly by assignment or subletting in response to changes in the company's marketing policy, whereas the portfolio of an institutional investor is much more difficult to change without disturbing confidence in the fund. Ideally, this should be changed by redirecting the annual inflow of funds rather than by a rapid restructuring of the stock of investment properties.

The portfolio may be physically diverse, a collection of properties meeting the client's functional needs in much the same way as plant or machinery. This must be distinguished from the compact urban estate which, at its most concentrated, is similar to an agricultural estate within its own "ring fence". Such estates in the private sector are normally of several hundred years standing, since they are now almost impossible to create without compulsory purchase powers. Very often they arose from the ownership of the land in its previous agricultural use prior to the urban growth of the 19th century, creating an opportunity for more than the aggregate management of a number of individual properties. The strict enforcement of covenants results in an environmental quality far higher than the most successful application of planning legislation by the local authority. Opportunities for marriage value will occur and in due course it may be feasible to carry out large-scale redevelopment.

Such an estate has a morphology consisting of phases of:

- pre-development
- development
- technical efficiency
- middle life
- old age
- refurbishment or redevelopment.

Judicious expenditure by the owner on regular maintenance, improvement or change of use may freeze or temporarily reverse any one of these stages. Local planning policy in general and listed building legislation in particular can significantly affect the life of individual buildings or the estate. Other legislation such as the Leasehold Reform Act can challenge the whole integrity of residential estates.

As redevelopment or refurbishment comes closer, considerable thought is required in order to obtain the best return from the existing estate without causing any problems in respect of the implementation and phasing of the works of renewal.

Finally, the portfolio must be managed in a manner consistent with the general ethos and role of the proprietary interests. There are many reasons, philanthropy, self-interest or pig-headedness, for not pursuing prudent estate management advice. It is the client's prerogative to disregard professional advice; it is the adviser's prerogative to give that advice, if necessary forcefully and frequently.

The property

Since the management of individual properties accounts for much the greater part of time spent by surveyors in practising estate management skills, these skills are the most highly developed. However, some property may be leased, some owner-occupied and some held for investment purposes.

Each of these requires a different approach to both overall management and detailed implementation of policy with the support of an appropriate records system. If required, some comparability can be achieved by making internal arrangements for owner-occupied property, such that the freehold is held within the group/parent, and operating companies or departments can be charged a market rent. A standard proforma for property records is unlikely to be able to capture, in a convenient way, all the necessary information for each of the three types of interests.

The relationship between the landlord and tenant (either directly or through their respective agents) has gradually changed from strictly contractual (and often adversarial) to one resembling more of a partnership between the parties. This change from a potentially litigious situation to a more friendly partnership seeks to be for the mutual benefit of both parties. Gordon Edington, in his book *Property Management — A Customer Focused Approach*, describes the adoption of such a philosophy to the management of properties owned by BAA plc but even that author has conceded that his approach cannot always be adopted for all properties and tenants. Whereas it is easy for a landlord to be generous and allow a tenant to terminate a lease when there is a constant demand, or even a waiting list, for such properties from potential tenants (as is generally the case with shops and offices at airports), any landlord, or his agent, who agreed to such a request for an industrial unit in an area of low demand might well just end up with a vacant unit.

Nevertheless, the principle holds good, provided it remains an equal partnership between the landlord and the tenant which serves both interests equally. Sometimes tenants may try to take advantage of such a philosophy and in these cases it may be necessary for the landlord to resort again to legal remedies, or at least the threat of them. This "iron fist in the velvet glove" approach is often adopted nowadays.

The occupier

Assuming that the greater part of management is carried out by or on behalf of landlords directly of occupying tenants, then the surveyors have in the past tended to create and perpetuate the myth that estate management can be depersonalised, in a way rather like that of the doctor trained not to get personally involved with his patient. However, this does not imply that there is not a real need to understand personal reactions, motivations and responses in order effectively to manage the landlord's interest. It is each individual tenant who is the real source of rental income and the quality of the covenant; property is merely the means by which we measure the quantum of the liability. By getting to know the tenant, the surveyor can see how successful he is in order to pre-empt rent arrears problems or anticipate the requirement for a larger unit.

The importance of tenant mix in achieving a balanced pattern of trading, and so a favourable investment yield on a shopping centre, is just one example of management of the occupier. It is not just a matter of implementation of user covenants but also of creating respect and mutual trust between landlord and tenant. As service charges become an increasing proportion of the cost of occupation, they have the capacity to use up much of the time of managing agents on relatively minor and occasionally frustrating disputes. The landlord's agent should look for a good working relationship with the tenant, particularly as in some circumstances the covenant requires him to make determinations between landlord and tenant.

In the field of housing, the occupier's needs are paramount. Housing is a basic human requirement; there is, therefore, an obligation on society to take decisions at the highest level on the standards to be achieved in housing and ensure that those standards are met. Public and voluntary sector housing are so distinct in character, specialist in function and detailed in content as to be subjects in their own right. The powerful interventionist role of statute in protecting the position of private residential tenants is now so complex that it has become a specialist area which is not covered in this book.

The users

Quite apart from the tenant, many other persons may properly be upon the premises: clients and customers specifically invited, trade representatives, employees of specialised service companies with responsibilities for the maintenance of the building or equipment and officers of central or local government with rights of entry.

Over a long period, common law, more recently partly overtaken by statute law, has created duties on landlord and tenant for the physical and environmental safety of such visitors to property. In addition, landlord or tenant may wish to influence the conduct or opinion of such visitors for property or trading purposes. The covered shopping centre creates a requirement for complex safety and security systems together with a wide range of services for the customers invited on to the premises. The same is true of transport interchanges or specialist recreational facilities such as historic houses with in excess of half a million visitors a year. Estate managers need to give more thought to the implications of their responsibilities as managers of facilities used by large numbers of the public.

These include:

- an excellent knowledge of the practical implications of the relevant law
- a well-trained security staff on the private property
- the use of technology
  - — closed-circuit TV
  - — video recording
  - — monitoring of all access points
  - — environmental monitoring
- regularly tested emergency procedures
- provision of suitable access (or a suitable alternative service) for those with disabilities.

The estate management process

The application of the general functions of management to urban estates can be seen as a process, the complexity of which, and frequency with which it is repeated, is appropriate to the nature of each individual estate. Lack of awareness of such a process does not mean that it is not occurring but that it may be occurring informally or in an ad hoc way. The smaller the estate the easier it will be to regard each decision as unique. That may make that decision easier but prevent the establishment of accepted procedures. The strong personalities of some proprietary interests may well feel threatened by the suggestion that each decision is not unique but capable of understanding and analysis within a structured process of management.

Strategy and tactics

A written statement of some broadly based goals will enable the policy-makers to give emphasis and weight between these and so react to changes in underlying factors. The problems facing long-term planning have already been identified, so the estate manager, having received the strategy and advised upon its formulation, should anticipate and encourage its regular review.

While there may be a unanimity of opinion by the estate owner, whether an individual, board or elected members, as to the goal or goals to be pursued, the estate manager has a more difficult task. Alternative tactics will exist, all of which can achieve that goal. For example, the balance of an investment portfolio can be changed in a variety of ways:

- change in pattern of new acquisitions
- disposals and new acquisitions
- refurbishment or redevelopment
- change of use of existing properties
- take-over of another portfolio
- exchange of properties between funds.

The time scale involved may be the crucial factor in selecting the most appropriate tactic, together with a perception of the response of interested parties. The choice of the preferred tactic tests all the qualities of the estate manager; judgment, knowledge, technical expertise, financial acumen and personnel management.

Some choices of tactics (from both the landlord and the tenant's points of view) for dealing with such matters as lease renewals, assignment, alterations etc are briefly summarised in Appendix B.

Implementation

This is the field in which the estate management staff will have had most training and experience. If the selection of the preferred tactic is followed by a clear and concise brief, and assuming staff of reasonable technical competence are available within an organisational structure appropriate for the task then implementation should be the least difficult part of the process. It can be assisted by the availability of an estate management manual and access to specialist advice.

The content and regular review of the manual is a major function of the senior estates staff; it is common in the larger property companies, partnerships and public sector departments. The manual will set out standard, statutory, contract and management procedures and identify sources of information for exceptions. The act of producing the manual and its regular review will provide senior staff with the opportunity to clarify their own approach. This textbook is not a manual, though it is hoped some of its content could assist senior staff in determining the type of manual and some of the content appropriate to their estate's needs. The sources from which specialist advice may be obtained can take many forms, but this is not just a matter of sources, it is rather one of adopting a positive approach to the problem-solving, and this is claimed to be one of the differences between the education of the graduate and those taking the professional examinations. Finally, the personal qualities of the staff involved must be stressed; considerable responsibility is devolved to individuals. Integrity and negotiating skill in discussion with those representing property interests and third parties can be developed through education and training but it must be said that to some it comes naturally and in others it is difficult to detect.

This implies a heavy responsibility on those in education to consider very carefully the real motivation of students whose entry to the profession is facilitated through university courses.

Control

While the information contained in a regular system of reports is interesting, it is only useful when measured against some objective, either quantitative or qualitative performance criteria. Since control follows from delegation, the pattern of responsibility will indicate the necessary form of control. Information must be collected and then presented in such a form that the decision-maker can reach conclusions on action at various levels of both policy and implementation.

Martin in *Shopping Centre Management* concludes that the most effective managers tend to be those who delegate (not abdicate) the administrative functions first and the human relations functions last.

The application of computer systems to estate management has changed the way estate managers go about their business. The computer and printer can issue rent demands, maintain clients accounts, allocate and apportion service charges, account for arrears, operate a forward diary, etc. Most modern computer program packages can also be integrated with other specialist systems for example to take the lease information and to use this to create property and even portfolio valuations and projections. The PISCES (Property Information Systems Common Exchange Standard) provides a common standard to enable this transfer of information.

Problems can be identified quicker and performance be analysed in a more rigorous way. Information for regular review by policy-makers is more readily available and more work can be handled by the same staff. Such systems do require a substantial commitment by the property management company and staff to ensure that proper training is carried out, backups are maintained and the system is updated when necessary.

To summarise, the first two chapters have shown the historical development of estate management and the application of the broad functions of management to urban estates. Of the many features of estates, their economic rationale is more fundamental than any other characteristics, therefore the following three chapters adopt this basis for their classification and more detailed study and analysis.

Sources and further reading

Adding value or adding cost, Jones, Management Services March 1994 p 20.

Property Management — A Customer Focussed Approach, Edington, Macmillan, 1997.

Quality:Total Customer Service, Taylor, 1992.

Shopping Centre Management, Martin, Spon, 1982.

3 Private Sector Estates

Nature of the estates

Estate management practice has developed within the private sector over several hundred years, but it is only over the last four decades that property has changed from a passive resource to a highly marketable commodity. Managed estates vary in size from Land Securities Trillium, with a portfolio of over 8400 properties, to that of the small residential landlord. Their extent may be as diverse as that of a supermarket with a branch in every high street or as concentrated as a high-value block in the City. Though they are diverse in terms of the type of property held and the legal status of the proprietary interest, there are a number of common features which unify private sector estate.

- The use of financial criteria in measuring performance and the preparation of annual reports and accounts that have to fulfill minimum requirements. The introduction of accounting standards and the asset regulations of the professional bodies have introduced a discipline into the treatment of the fixed assets in these accounts.
- Accountability to a defined group, in a company the shareholders, though in practice this may be one or two major institutions.
- Either liability to tax in some form, on income and capital, or exemption from tax, both situations significantly affecting the acquisitions policy and portfolio management.

Property may be the *raison d'être* of a private sector organisation or just one of a number of resources which has to be marshalled and co-ordinated to carry out the trading activity. In a property company the management of the portfolio is the primary, indeed sole, activity. Though the objectives may be investment, development and/or trading.

A feature of the direct investment market over the past decade has been the increasing number of stakeholders and vehicles involved in the market. These range from small funds, private equity vehicles, overseas investors and funds combining private and institutional investors through to high net worth individuals using heavily leveraged vehicles to acquire single property assets and small portfolios.

The three main types of private sector owners discussed here are institutional investors — insurance companies, pension funds; property companies; and occupiers of property in commerce and industry. These three groups have different motivations that drive their holdings of property.

- In an institution, property competes with other asset classes to provide returns consistent with the objectives of the fund. For directly held property at least, the main value of property in a fund is that of a diversifier versus the performance of equities particularly.
- In a property company, property is the main focus of the company, either through the development of assets for sale or the trading of property to generate returns to shareholders.
- In commerce and industry, either as owners or as occupiers of an institution's investment portfolio, property is an asset to be utilised as efficiently as possible to support the core trading requirements of the business. Increased focus upon asset value within corporate accounts has raised awareness of the importance of property on the balance sheet.

An estimated 50% of all commercial space is owner-occupied, the remainder is held by institutional investors, traditional landowners, overseas investors and property companies.

Of the remainder, institutional investors — insurance companies, pension funds and unit trusts — collectively hold some 40% of commercial property in the UK. An estimated 36% is owned by property companies — listed and private. Overseas investors hold circa 15% and around 5% is held by traditional landowners like the Crown, the Church and the University Colleges. Limited partnerships, the rising stars in the investment firmament hold about 4% having come from nowhere over the past five years.

Table 3.1 Estimated 2004 value of the UK Investment market for commercial property

	£bn
Institutions	101.6
Property companies	91.4
Overseas investors	38.1
Traditional landowners	12.7
Limited partnerships	10.2

Source: IPD, ONS, Jones Lang LaSalle

It is important to differentiate between those who own property purely for investment purposes and those who invest in property as a by product of owner occupation. Although their objectives may have similarities at the top level, there is a marked difference in emphasis. Investors will be driven by the metrics of portfolio and asset management, with property management as a maintenance function. Owner occupiers will be driven primarily by the operational functionality of the building, with the investment return being of secondary importance.

Over the last 20 years there have been many changes to the structure and nature of property ownership and investment that affect all types of private sector estate owners.

- The regulatory environment that governs investment has changed significantly. These changes impact on property investment directly and indirectly. Among the indirect effects are the Basel II New Accord on bank lending, solvency and risk capital, which introduces a new regime for banks to manage risk and meet liabilities. This has obvious implications for the property debt market.

  Changes to international accounting standards have an impact on the property market, with new structures for reporting operating and finance leases set out in IAS17. Combined with proposals to bring operating leases on balance sheet as assets and liabilities, there may be major changes to firms' accounting and financial ratios.

  The Finance Act 2003 replaced Stamp Duty with Stamp Duty Land Tax (SDLT). SDLT is a tax on the substance of the transaction

Table 3.2 Typical performance objectives

Fund	*2001 size £m*	*Specialisation*	*Performance objective*
Norwich Property Trust	560		To outperform the IPD Monthly Index by 0.5% over a rolling three year period
Igloo Urban Regeneration Fund	200	Urban regeneration projects	to produce an ungeared project IRR of at least 1.5% pa over the life of the fund
Prudential M&G Pooled Property Fund	327		To outperform the average of competing pooled funds as measured by the CAPS survey of UK Pooled Pension Funds
Corporate Centres Partnership	20	Serviced offices	To outperform the IPD Office Series by 2% pa over 10 years
NU Public Private Partnership	50	PFI projects	To produce a real IRR of 6.5% pa over rolling six year periods
CGU General Fund	205		To outperform the IPD Monthly Index by 50 basis points pa on an annual rolling basis
Henderson UK Property Fund	182		To be in the upper quartile of all property funds measured in the CAPS Pooled Property Universe on a rolling five year basis

rather than on the instrument effecting the transaction. It is therefore a tax on land transactions, being any acquisition of a chargeable interest, other than an exempt transaction. SDLT rates were structured to reflect the Stamp Duty regime being replaced so in the case of property acquisition very little difference in duty has been seen. In the case of leasing however, the net effect of SDLT has been to increase transaction costs substantially.

- The outsourcing of the estates management function to specialist asset managers has become common practice in all types of private sector estate. This facilitates the operation of targeting and measurement, with fee levels often dependant upon the achievement of explicit performance targets over time. Table 3.2 shows typical performance criteria.

- On the back of the outsourcing wave an asset management industry has grown up providing specialist expertise.
- The pattern of ownership of properties has changed. In particular the globalisation of business has led to overseas ownership becoming far more common, adding different reporting practices, legal systems and cultures to the mix.
- Methods used to fund and finance property have changed over the last two decades. Conventional bank-led project finance followed by institutional take out or long term mortgage funding has been largely replaced by more complex mixes of debt and equity and spreading of risk through syndication and securitisation.
- Analysis of performance has become more rigorous and standardised. The pooling of property data by institutional investors particularly in the IPD has ensured that a detailed performance benchmark is available to the industry. Operational Property Databank (OPD) perform the same function for occupiers of property but using operational parameters rather than financial ones.
- Indirect property investment vehicles have increased their market share substantially. The UK private property vehicle market has seen rapid growth to reach a market capitalisation of £23 billion in 2001. Limited partnerships alone have grown over the period 1996 to 2001 from just over £1 billion gross assets to over £13 billion. Major features of this market over the last five years include a growth in pan-European vehicles and, in particular, a shift in investment style from relatively conservative core funds to more aggressive, highly geared core plus and opportunity funds.

The institutions

Institutional investment in property was in decline as a proportion of the fund holdings between the early eighties and late nineties as its positive characteristics as an asset class were submerged beneath a wave of superior performance by equities.

In the noughties, however, property has staged a modest recovery in popularity and the proportion of a portfolio held in property is rising again. Throughout the period the value of property actually held by institutions grew substantially.

Table 3.3 Insurance company property assets 1981–2003

	£m	Property as % of overall assets
1981	15,940	21.5%
1982	17,561	18.1%
1983	18,809	16.3%
1984	20,766	15.3%
1985	22,181	14.4%
1986	24,140	12.8%
1987	29,305	14.3%
1988	36,756	15.6%
1989	42,629	14.7%
1990	38,431	14.0%
1991	35,470	11.0%
1992	32,781	8.7%
1993	36,538	7.5%
1994	38,275	8.3%
1995	37,92	6.7%
1996	38,520	6.0%
1997	45,364	5.9%
1998	47,408	5.4%
1999	51,761	5.0%
2000	52,794	5.1%
2001	55,088	5.4%
2002	53,622	5.5%
2003	58,221	5.7%

Source: IPD

Table 3.3 shows the historic pattern of decline from the perspective of insurance companies. Pension funds show a similar pattern with property holdings having recovered slightly to 7% of fund value in 2003.

The characteristics of any particular portfolio are tailored to meet the performance criteria set for the fund. The last 20 years have seen a major change in the structure of institutional portfolios. Although the proportion of funds held in property has declined, this is in part a function of the growth in value of other investments. In fact institutional investment in property has grown every year since 1980.

In 2003, IPD figures showed £2.5 billion net disinvestments by institutions — the first time this had been observed since 1980.

The data shows three main trends.

- A substantial diversification away from offices — mainly Central London offices, which accounted for 32% of total capital value at the end of 1981 and have tended to be one of the most volatile parts of the UK market.
- Increased exposure to retail property, which has been one of the better performing sectors and has tended to produce relatively stable returns. Retails' share of total portfolio value has climbed almost continuously from 27.2% at the end of 1981 to 44.0% 20 years later.
- Increased exposure to new property types. Within each sector, institutions have increased their holdings of new property types such as retail warehouses, office parks and distribution warehouses. Collectively, these three types accounted for 21% of total capital value at the end of 2001.

The balance of the portfolio and its performance is developed further in Chapter 10, but possibly more important than the balance of properties making up the portfolio is the positive management of the portfolio, continually seeking out opportunities for increasing the return on the capital invested.

From a management standpoint the characteristics of each sector are subtly different. Offices tend to be larger tenancies with, depending upon the age and location of the premises, less of a management overhead. Retail management typically involves more, smaller tenancies with a relatively high overhead looking after the more onerous public liability aspects of shops. Industrial management will have larger tenancies, but a different set of management problems with respect to security and environmental impact.

Within the broad category of institutions, further subdivision by type of fund is becoming increasingly complex as mergers, acquisitions and outsourcing of asset management abound. Many insurance companies have outsourced asset management on a discretionary basis, often to a group subsidiary which then seeks to attract additional third party business. Pension funds have gone down the same route making statistical distinctions between approaches difficult. Inevitably, therefore, analysis by institution type is presenting a different facet of the same physical investment stock.

Table 3.4 Holdings of pension funds and long term insurance funds by sector and sub-sector 2001

	%
Retail	
Standard shops	14
Shopping centres	5
Retail warehouses	12
Other retail	3
Offices	
Standard offices	32
Office parks	6
Industrials	
Standard industrials	9
Industrial parks	3
Distribution warehouses	3
Other property	3

Source: IPD

Table 3.5 Changes in weighting between sub-sectors 1981–2001

		% of capital value		*Change %*
	1981	*1991*	*2001*	*1981–2001*
Standard shops	13.5	17.3	13.5	0
Shopping centres	10.5	14.5	16.3	5.8
Retail warehouses	0.4	3.1	11.6	11.2
Other retail	2.8	2.7	2.6	–0.2
Standard offices	54.9	44.6	32.7	-–22.3
Office parks	0	2.6	5.9	5.9
Standard industrials	12.6	10	8.7	–3.9
Industrial parks	1.1	2.3	2.6	1.5
Distribution warehouses	0.5	1.1	3.4	2.9
Other property	3.6	1.8	2.7	–0.9

Source: IPD

Table 3.6 Insurance companies and property holdings

£ billion 2004	*Funds under management*	*Direct property holdings*
Prudential	126	14.8
Aviva	240	10.0
Standard Life	94	9.1
Legal & Geneal	103	6.5
HBOS	77	5.4
Lloyds TSB	108	4.0

Source: RETRI Group

Insurance companies

Apart from a few specialist companies, insurance companies are involved in a wide range of business interests. A substantial part of their business is in long term liabilities and this is one of the justifications for property investment. The largest ten companies account for 72% of the long term market. Some of the largest companies are identified in Table 3.6 with information on the size of total funds in 2004.

Given the high value of the assets under their control, insurance companies have always been able to buy enough real estate to mitigate any risks inherent in a lack of diversification between different property sectors.

To the extent that they hold shares in property companies and property is part of the asset base of all their other shareholdings, institutions have rather wider property interests than is suggested by the figures for their direct property portfolios.

Pension funds

Pension funds are substantially more risk averse than other types of institution. They began to invest in property in the late 1960s and increased their exposure heavily in the 1970s. A relaxation of the rules governing the holding of overseas equities plus the newly available index linked gilts saw the trend into property reversed in the 1980s. This trend saw property holdings in pension funds decline from over 20% in 1980 to a low point of 4.5% in 1999. However, there is some

Table 3.7 Pension funds by size 2004

	Fund size £ billion
British Telecom	22.8
Universities Superannuation Scheme	19.3
Coal Pension Trustees	18.8
Electricity Pensions Services	15.1
BP	12.5
Scottish Public Pensions Management	12.0
Royal Mail Pensions Trustees	12.0
Railways Pensions Management	11.8
National Grid Transco	11.0
Shell Pensions Management Services	10.7

Source: Pension Funds and their advisors

evidence that property is becoming popular again post the millennium and allocations have increased slightly over the past few years.

Table 3.7 shows the top 10 UK pension funds by size. As can be seen, this end of the size spectrum is dominated by the pension funds of the public and privatised utilities, with only the oil companies breaking up the pattern. The largest pension fund of all — British Telecom has nearly £23 billion of assets of which just over 14% are invested in property.

This, and the Universities Superannuation Scheme with just over 11% in property, are the exception among the top 10 funds. The Royal Mail and Railways funds see 8% and 7% property investment respectively, but the average allocation stands at just under 6%. The top 10 pension funds account for just under £11 billion of property investment. As far as pension funds are concerned, the ownership of property seems to be a function of scale. Table 3.8 shows a breakdown of asset ownership by fund size.

Clearly, the bigger funds hold a higher proportion of their assets in direct property than do the smaller ones. Given the large lot sizes typical of property as an asset class, smaller pension funds often find it difficult to gain sufficient diversification. In time this may be resolved in part by a securitised, tax transparent vehicle such as is being proposed by the UK Treasury.

Table 3.8 Asset allocations by fund size, 2003

	0–5	*5–10*	*10–20*	*20–50*	*50–100*	*100–250*	*250–1000*	*1000+*
UK equities	52.5	53.3	52.3	50.8	49.3	48.5	44.7	45.0
Overseas equities	21.5	17.1	18.9	19.7	19.7	21.1	21.7	20.8
UK fixed interest	13.6	12.3	13.0	12.8	14.2	11.9	14.1	12.0
Overseas fixed interest	1.8	2.2	1.8	2.0	2.5	2.0	2.4	3.1
Index linked gilts	3.5	7.5	4.7	4.6	4.1	6.3	7.7	7.8
UK property	1.7	1.7	1.9	1.3	1.6	1.7	2.7	5.8
Overseas property	0.0	0.0	0.0	0.0	0.2	0.0	0.0	0.1
Cash	3.4	3.8	3.5	3.7	3.2	2.7	2.8	1.0

Source: Pension funds and their advisors

Asset managers

As at June 2003 the UK Asset Management sector managed £2 trillion of funds in the UK and a further £5 trillion overseas. The five largest companies asset managers hold 28% of the market. Nearly 40% of UK managed assets are controlled by groups belonging to insurance companies. Investment banks control 19%, retail banks 18% and the rest are independent.

Nearly 40% of the assets managed by the industry are owned by insurance companies and 37% by pension funds. Table 3.9 shows the top 10 asset managers ranked by the size of funds under their management, together with their owners. A survey taken by the Investment Management Association (IMA) for 2003 among firms with over £1 trillion under management showed UK property to be a relatively minor asset class.

Table 3.10 shows significant differences between the assets managed for insurance companies and pension funds, traditionally the twin bastions of the institutional market. Overall, property accounted for 5.5% of the asset allocation.

Table 3.9 Asset managers by funds under management

	£bn	Group
Legal & General Investment Manageement	123	Legal & General
Barclays Global Investors	119	Barclays
M&G Investment Management	116	Prudential
Morley Fund Management	104	AVIVA
Standard Life Investments	78	Standard Life
Scottish Widows Investment Parntership	73	Lloyds TSB
DWS Investments	70	Deutsche Bank
Henderson Global Investors	69	AMP
Insight Investment	68	HBOS
Schroders plc	67	Schroder

Source: IMA

Table 3.10 Asset allocation by UK asset management companies 2003

	UK equities %	*Overseas equities* %	*Gilts* %	*Bonds* %	*Cash* %	*Property* %	*Other* %
Pension funds	32.9	22.5	21.2	11.3	3.7	1.5	6.9
Insurance companies	22.7	8.0	25.8	22.9	8.6	8.6	3.4
Retail funds	37.4	33.4	6.2	8.7	7.4	0.5	6.4
Other	12.3	24.6	16.3	19.7	10.4	9.4	7.2

Source: IMA

Property Unit Trusts (PUTs)

Property unit trusts can be either Authorised or Unauthorised. Authorised property unit trusts are one of the few vehicles open to retail investors in property. Units in them trade at asset value. They have an open-ended structure, which means that the fund changes in size as investors enter (buy units) or leave (sell units). Because of liquidity constraints these funds tend to retain quite high levels of cash and also invest in property company securities.

Table 3.11 UK Property Unit Trust values

December each year *£ billion*	*Fund value*
1998	5.6
1999	6.1
2000	6.4
2001	7.3
2002	8.0
2003	10.3
2004	14.1

Source: APUT

Unauthorised property unit trusts exist mainly to provide an efficient investment conduit for tax exempt investors, such as pension funds. This unit trust structure is also utilised as a feeder fund for other tax exempt investors, such as SIPPs, to invest in vehicles like limited partnerships.

Following liquidity problems in the early nineties, PUTs have gone from strength to strength, almost trebling their fund value between 1998 and 2004.

Other indirect vehicles

Private Property Vehicles (PPVs) are collective investment schemes that are not listed. The European PPV market has seen rapid growth in recent years. At the end of 2004 there were over 300 such vehicles in Europe with a Gross Asset Value approaching Ä250 billion.

By far the most popular domicile for PPVs by number has been the UK. The UK domiciled vehicles mainly invest in the UK and are typically limited partnerships, property unit trusts and managed funds. However, the Finance Act 2004 introduced a basis for charging SDLT on transfers of land by a partner into a partnership or from a partnership to a partner and on the transfer of an interest in a partnership. Whether this tax penalty is enough to force limited partnerships offshore remains to be seen. In any event they are planned to be joined by a completely new structure — the Property Investment Fund (PIF), a REIT for the UK.

The introduction of PIFs was one of the proposals emerging from the Barker Review of Housing Supply announced in the interim report published in 2003. With the March 2005 budget, the Government issued a further discussion paper on the proposed UK property collective investment vehicle. Subject to resolving outstanding issues, the Government aims to legislate for the new vehicle in the 2006 Finance Act. This means that the new vehicle may be available from around mid 2006. The approach is for a flexible vehicle with few restrictions on types of activities, properties and minimum holding period.

The Government has decided that PIFs will be allowed to undertake property development, and there will be no restrictions on type of property, property sector or location. PIFs will be allowed to invest in property located anywhere in the world. There will be no requirement to hold a minimum percentage of residential property and REITs will be allowed to be internally or externally managed.

There will be a requirement for 75% of the PIF's gross income and assets to be derived from property letting. The PIF will be required to distribute 95% of its net property letting income to investors (after deduction of capital allowances). Income and gains from the property letting activity will be exempt from corporation tax in the PIF, but will be taxable as normal property income when distributed to the investor.

The favoured structure for PIFs is as close ended companies held by a wide number of shareholders. The Government has not yet decided whether there should be a requirement for PIFs to be listed but the tax advantages of such a vehicle could well cause some listed property companies to change their status.

Property companies

At the end of 2003, the FTSE index contained 53 companies listed as "real estate holding and development". They had a combined market capitalisation of £24.6 billion — under 2% of the overall market.

Since the late 1980s, the property company sector has performed relatively strongly compared to the equity market as a whole. Over the five years to December 2003, property companies delivered returns of 4.5% pa, compared to just 1.1% for all shares. Over the same period, however, Jones Lang LaSalle (JLL) figures suggest 11% returns in property.

In the late 1990s and into 2000, the property sector under performed the overall stock market. The even larger discounts to Net Asset Value that ensued (traditionally property companies have been valued at a substantial discount) led to a number of companies delisting, notably MEPC, with many others buying back stock and increasing their private capital.

In addition to the withdrawal of some property companies from the market, the last decade has also seen something of a polarisation of strategic direction amongst some property companies.

A number of firms have switched from asset ownership to asset management — setting up specialised funds for external investors. An example of this approach would be Pillar Property. As property adviser to both the Hercules Unit Trust, the largest retail park investment trust in the UK, CLOUT, which specialises in City of London offices and PREF, a european retail park fund, Pillar asset manages property with a gross value of over £2.4 billion.

Pillar receives management fees on the trust portfolios and performance fees dependent upon trust investment performance exceeding certain benchmarks. In addition, Pillar develops retail and schemes which will be offered to Hercules. The majority of Pillar's investments are in the form of units in Hercules Unit Trust of which at March 2004 Pillar owned 35.29% and in the City of London Office Unit Trust of which Pillar owned 36%. Pillar owns 49% of PREF.

Others, such as Land Securities Trillium, have taken the management concept further to provide a complete outsourcing product. It provides a complete property outsourcing package, combining:

- property financing
- facilities services
- capital projects delivery
- estate strategy and asset management
- provision of new space (property acquisition and development).

The solution enables an organisation to transfer its short term and long term property needs to a single specialist provider. Trillium takes on the ownership, management and development of all or part of a client's estate, enabling property assets and liabilities to be converted into an integrated property service. This releases capital for the client, delivers operational savings, provides occupational flexibility, reduces property risk and provides price predictability.

At the other end of the spectrum some companies focus upon the high added value (and higher risk profile) that can be achieved through the development process. In some cases these companies have been conceived to provide occupational or investment stock — examples would be Gazeley (owned by WalMart) and BAA Lynton.

Elsewhere, there has been a switch towards a more entrepreneurial approach: for example, taking completed and let developments then securitising them using the proceeds for new activities or returning capital to shareholders.

Overall property companies have begun to see their business as providing business services, be these premises or management or both.

Annual reports

The annual reports of the publicly quoted property companies demonstrate a varying degree of disclosure to shareholders and interested parties. In addition to the statutory requirements, portfolio analysis by user, tenure, geography and value is appropriate, together with lists of major properties.

The chairman's report and the notes to the accounts offer opportunity for comment on key issues, such as the application of depreciation concepts, which have a more fundamental effect on property companies than other types of companies. The British Property Federation exists to enable the companies to lobby with one voice on matters which affect them generally such as depreciation and the widening impact of professional, national and EC guidance/directive on company and property matters.

On 7 June 2002, the European Parliament approved legislation requiring all listed companies in the European Union to prepare consolidated financial statements under EU adopted International Financial Reporting Standards (IFRS) for financial years beginning on or after 1 January 2005.

A fundamental concept underpinning IFRS is that of fair value. For the property sector the international standards dealing with investment properties, taxation and financial instruments cover the essential requirements that reflect this principle.

- IAS 40 — Investment property encourages users to hold investment properties at fair value but gives the option of using depreciated cost because of the problems in obtaining reliable valuations in some markets.
- The standard dealing with tax (IAS 12) adopts a balance sheet

approach. Timing differences between the carrying value of an asset and its tax base result in a deferred tax liability or asset. For investment properties, this means that under IFRS an unrealised capital gain (or loss) that derives from the annual valuation of a property will give rise to a deferred tax liability (or asset) that is not currently recognised under UK GAAP until there is a binding sales contract.

- The new standard makes a clear distinction between operating and finance leases. Finance leases are those that transfer substantially all the risks and rewards incident to ownership to the lessee. All other leases are operating leases.

Under the revised IAS 17 issued in December 2003, land and buildings elements must be classified separately. The land element will normally be an operating lease (unless title passes to the lessee at the end of the lease term) and the rent handled on a straight-line basis.

The buildings element could be treated as a finance lease. This could happen, for example, where the lease term is for the major part of the economic life of the asset or if, at the inception of the lease, the present value of the minimum lease payments amount to most of the fair value of the leased asset. If the buildings element is treated as a finance lease, a matching asset and liability will appear on the lessee's balance sheet — calculated as the lower of fair value and the present value of the minimum lease payments attributable to the buildings. Although the net effect will be zero, the company's gearing will be affected.

Clearly the classification of leases could become crucial to the apparent financial health of a company. The BPF publish guidelines in the form of a checklist to assist with the process.

Property company analysis

While the substantial aim of the text is the effective management of the individual properties in a portfolio, it would be wrong not to recognise that property companies are part of the equity market. Hence, the proprietary rights in the aggregate portfolio, in the form of shares traded, is in the province of the stock brokers and their analysts. The distinguished financial journalist Michael Brett has considered this interface in detail in *Valuation and Investment Appraisal*, edited by Clive Darlow, and Milner has taken an accountant's view in *Property Company Accounts*.

It is important to note that property shares tend to have a high PE ratio compared with other specialist equity sectors. In considering PE ratios, price is the markets reaction to events, whereas earnings (on an agreed basis) are closer to an accounting fact. Care is needed where earnings are influenced by an exceptional factor and share price is influenced more by expectations as to future earnings rather than historic accounts. In property valuation terms, PE ratio is a similar concept to YP.

The yield on a share (dividend/share price) is a similar concept to yield on a property (net rental income/capital value). The problem is that the route from the company's profit through to the dividend actually paid can be tortuous. As a result from year to year the dividend may be greater than, or less than the profits.

Property companies are perhaps unique in that they are backed by numerous assets which are of a tangible nature. Each asset is capable of being sold separately, and after deferred contingent tax payments as necessary, the aggregate net proceeds will (in theory) exceed the market capitalisation.

An analysis of any type of general trading company is incomplete unless a check is made upon it to establish whether beneath the figures there may be a latent and dormant potential property company. That is to say under performing freehold and long leasehold property assets, which could be divorced from their operational user, without destroying the operational users activity as a viable general trading activity — in short, what was once, perhaps unfairly, called asset stripping.

There are many accounting ratios which are calculated directly from the standard information found in the annual accounts of all companies. Their appropriateness and importance varies between the different specialist equity markets. One of the most important for property companies is the ratio of resources to debt, which may be considered in capital or revenue terms. Perhaps the simplest to consider in the balance sheet is capital valuation:debt ratio; if less than one, then bankruptcy seems inevitable! Typically, the ratio of debt:shareholders funds is a measure of the gearing, indicating the extent of the exposure to one kind of risk.

The same principle may have more immediate effect on the profit and loss account, where the annual equivalents of debt and capital value, that is to say interest payments and net income respectively, can be compared.

It must be clearly understood that the relationship between capital and annual gearing is not symmetrical. Since variations in interest rate affecting variable rate loans or short term fixed interest loans result in changes in the annual interest payment, even though the capital debt is unchanged. Also, the refurbishment of a portfolio may reduce income due to voids and hence worsen the ratio, but this may be only a temporary, and welcome, indication of the improving quality of the underlying assets.

Case studies

British Land and Land Securities are the two largest property companies in the UK. Both have similarly sized and structured portfolios of property

£bn 2004	*British Land*	*Land Securities*
Investment properties	9.25	8.15
Net debt	4.87	2.44
Property income	0.57	1.48

£bn value 2004	*British Land*	*Land Securities*
Retail	5.53	3.44
Office	4.57	2.44
Other	0.54	0.35

Source:

The significance of a few properties is apparent in the case of British Land where Broadgate accounts for 25% of the total value of the portfolio.

The Grosvenor Estate illustrates the diversity of activities found within property companies. Grosvenor is owned by trusts for the members of the Grosvenor family. The origins of the property business can be traced back to 1677 when the areas of London, now called Mayfair and Belgravia came into the ownership of the Grosvenor family. Development of Mayfair began in the 1720s and Grosvenor has been involved in the development and asset management of property ever since.

Table 3.12 Portfolio analysis by region 2003 — Grosvenor Estate

2003	*£m*
Continental Europe	980
Australia/Asia Pacific	231
UK/Ireland	4,061
Americas	782

Source: Grosvenor Estate

Grosvenor is active across many sectors of the property industry — offices, business parks, shopping centres and residential. A particular specialisation is that of complex urban projects where all the uses are mixed together with long term commitment and an ability to work with local communities to ensure that the regeneration of their neighbourhood meets their wishes. Grosvenor currently manages a portfolio of some £5.4 billion and has interests in 16 countries.

Commercial space, in offices, business and industrial parks is the most substantial category within Grosvenor's portfolio, representing some 51% of assets under management. Grosvenor's interests range from high specification new developments in the City of London, to business parks in Silicon Valley, California, high rise buildings in Hong Kong and 18th century period office buildings in Mayfair

Retail, both in shopping centres and high streets represents some 32% of assets under management in 2004. Grosvenor is active in the development of shopping centres in each of its regional operating areas and retains an investment in many of the centres it has developed. In 2004 Grosvenor had interests in 43 shopping and retail centres across its four operating regions with an approximate gross lettable area of 1,514,785 m^2.

Grosvenor has interests in residential property across each of its international operating regions, however, the majority and best known residential parts of the portfolio are in London, in Mayfair and Belgravia. The residential sector accounts for 17% of Grosvenor's assets under management and 60% of these are in London including internationally renowned addresses such as Eaton Square. As well as being a long term investor, Grosvenor is an active developer of new residential property in all its operating regions. Many of Grosvenor's new developments are carried out as joint ventures such as the creation of five new town houses at Dorset Mews in Belgravia with

Mountcity London Residential Ltd or 175 luxury apartments at Repulse Bay, Hong Kong, with the Asia Standard Group.

In addition to the developing and owning of property Grosvenor is also a substantial fund management business with over £2.7 billion in third party assets under management. Some of its specialist funds are listed below.

- GMETRO — a central London office fund set up in 1999. It was created to take advantage of the investment opportunities in London and its status as a leading international location for business and financial services.
- The Grosvenor Festival Place Fund — a 10 year limited partnership fund created to manage the Festival Place retail development, which was completed in October 2002.
- The Grosvenor Land Property Fund — A 50/50 joint venture with Hong Kong Land.
- The Grosvenor Shopping Centre Fund — set up in 1998 as a 10 year limited partnership created to invest in prime, dominant shopping centres around the UK.
- YK Fiduciary Fund — A joint venture between Grosvenor, Ayala International and Uni-Asia Finance Corporation.

Occupiers

Property is a unique corporate resource. On the one hand, it is an asset that needs funding, on the other, it is a factor of production. Usually, it is the largest asset in an organisation and, after staff costs, the largest expense item.

Traditionally, corporate real estate has been thought of as a tangible asset of an organisation. The trend has been to view this tangible asset as illiquid, under utilised and often under performing, particularly in terms of organisational cost of capital. Taking a more holistic view it can be seen that corporate property also has some important intangible assets that have an important impact upon its function in the organisation.

The concept of intangible asset is based on the premise that sustainable value can be created from developing aspects such as the skills and knowledge of the workforce, the information technology that supports the workforce, the organisational climate and the physical workplace.

Most of the expenses associated with property are treated as cost items, thus the generally-accepted view of corporate real estate is as an overhead to be minimised. However, in recent years the property and facilities management industries have begun to shift their focus from proving their worth by saving money, to demonstrating that corporate real estate adds value to an organisation. Much of that added value is demonstrable in the interaction between the buildings and the workforce.

Most UK businesses are small or medium in size. In smaller companies the finance director or company secretary is usually responsible for property matters, from time to time instructing surveyors on an *ad hoc* basis to advise or act in respect of property matters. Typically, property will not be discussed by the board or management team until an accommodation problem arises.

In larger companies, perhaps with multiple sites, property and facilities managers are employed to address the day to day operation of the estate. However, senior management often overlooks property strategy until there is a conflict between business needs and property provision.

In large companies, particularly multiple retailers, a complete organisation can be formed consisting of architects, quantity surveyors, property managers and other specialist staff, creating a multi-disciplinary team able to design layouts and bring into use new units in the least possible time within the fast moving retail market.

Surveyors in commerce and industry tend to deal with specialised type of property but require a breadth of knowledge in order to assess the impact of the whole range of factors which affect their premises. In acting for a single client, in aggregate they represent many of the tenants who occupy the investment portfolios of the institutions, hence their perception of property as a cost to their business is very different from that of the insurance companies and pension funds.

The tasks

Traditionally, the two main activities of the in-house estates department were acquisition/disposal and day to day management of facilities. Increasingly, however, the estates department is becoming involved with the overall occupation strategy from location through to efficient occupation of space.

The programme of acquisitions, or in some cases development, may be derived from the corporate trading strategy. This is extremely important for companies in direct contact with the public, where site selection influences turnover and physical appearance contributes to the company's identity in the eyes of the public.

The property department's knowledge of the property market and alternative means of financing acquisitions and development will make a significant contribution to the overall profitability of the company. The property itself may be owner-occupied or rented and within a large company there will inevitably be some lettings of accommodation retained or sublet for various reasons.

Much of the regular management work will consist of rent reviews, renewal of leases, maintenance, rating and the growing number of administrative responsibilities attached to the occupation of property.

Company annual reports include substantial information on property; the accounting standards on current cost accounting and depreciation together with the RICS statements of asset valuation practice represent a major task for the property manager, who will be concerned to obtain the highest quality of mandatory and discretionary property information provided in the annual report.

The means by which the estate management function can make a proper contribution to corporate management will depend partly on the formal pattern of responsibility between senior executives and directors and partly on the way in which property works for the company. In some companies, the property is held in the group and internally charged market rents enable the management to see each site as a profit centre.

In-house or consultants?

It is unusual to find an estates department which carries out every aspect of the management of the company's property. The reason for this will be partly the way in which the company and its estates department have developed over the years, and there is seldom a conscious review of the allocation of tasks within and outside the company. However, there is some general agreement on the factors which result in the employment of consultant surveyors, either on an ad hoc basis or in a retained capacity.

- The geographical spread of the portfolio may be such that it would be financially inefficient to deal with some or all the properties from one office.

- Certain specialist skills may be required infrequently and it is unrealistic to expect staff to have that expertise.
- The cost of obtaining and keeping up to date the range and quality of certain types of market evidence and information may be beyond the scope of an in-house estates department. In certain circumstances an external report or valuation may be better for the particular task that has to be undertaken. Strictly, it may appear better in the eyes of third parties, whether it is intrinsically better or worse is another matter.
- The quantity of work involved, or the speed with which it must be done, may be quite outside the capacity of the size of the establishment.
- A commercial market fee basis, which may be used to price the "value added" of the estates department.

The estate manager will therefore be closely involved in agreeing on some general policy on the selection and appointment of external surveyors.

Case studies

Given the sheer diversity of the potential occupier base it is not possible to explore individual property strategies for each occupational type. The following case studies represent the range of corporate property solutions present.

British Telecom

As can be imagined British Telecom has a massive portfolio of operational property. Over the years the company was keenly aware of the cost burden that this imposed on the company and managed to reduce the costs of its estate by £500 million between 1990 and 2000. Since then BT has been very innovative in the use of property, particularly the development of Work Style 2000 — flexible office accommodation to promote and alternative workplace strategies.

The liberalisation of telecommunications in the UK and the need to compete for business in a market stacked against the incumbent by legislation; the need to invest substantial sums in new technologies like broadband; the purchase of third generation mobile phone licences and investment hungry expansion all put significant pressure on the availability of capital. By the end of 2000 debt was approaching £25 billion. This increasing debt coincided with a drop in the share price of 50%

In November 2001 contracts were exchanged with Telereal, a 50:50 joint venture between Land Securities Trillium, a subsidiary of Land Securities, and the Pears Group to sell the entire portfolio and lease back the facilities.

The properties transferred comprise offices, telephone exchanges, vehicle transport depots, warehouses, call centres and computer centres. Telephone exchanges and related technical buildings account for 75% of the total, with offices and other properties constituting the remaining 25%. Freehold or valuable leasehold with 5% short leasehold made up 95% of the properties. The BT Tower and BT Centre in Newgate Street were excluded from the transaction.

The transaction included the transfer of approximately 350 employees from the BT in-house corporate property team to Land Securities Trillium Telecom Services Ltd. Under the terms of the arrangement, LST Telecom meets all the employment obligations and recovers the costs associated with this arrangement on a fully indemnified basis from Telereal.

The deal covered nearly 7000 properties with a total floorspace of 5.5 million m^2 and a value of around £2 billion. The divestment of the property estate will enable a more flexible approach to BT's office arrangements and building requirements.

Tesco

In 1998 Tesco set out its four part growth strategy focusing on core UK business, non food, retailing services and international expansion. Since then the company have moved from being the number three domestic retailer to being one of the top three international retailers in the world with 2318 stores, 326,000 people and 4.2 million m^2 of selling space. In 2004 turnover in the UK was £24.8 billion from a sales platform of 2.2 million m^2 in 1878 stores. Through expansion of retail services Tesco has 5.2 million banking, Tesco.com and telecoms accounts. Tesco.com is the world's biggest online supermarket. Tesco Personal Finance alone achieved profits of £160 million of which Tesco's share is £80 million.

Notwithstanding the scale of Tescos operation, from a property standpoint particularly the rate of growth has been ferocious. The number of stores in the UK alone has almost trebled since 2000 through development and acquisition. To support this growth requires cutting edge distribution warehousing.

By way of example the Tesco ambient food regional distribution centre is located at Thurrock in Essex. Constructed within a disused chalk quarry, the 46,000 m^2 development took less than 18 months to build, and led to the award of a £60 million long term logistics management contract to logistics company Tibbett & Britten. The building handles up to 2.5 million cases of food products a week, contains 3.2 km of aisles and 64 loading docks and employs 1,100 people.

Operating 24 hours a day all year round, the new centre is designed to cater for the continuing growth of Tesco in the region, and is one of the most technically advanced in the UK. It can store up to 45,000 pallets of food and beverage up to five or six high, has 7500 picking slots, and is capable of handling 130 million cases a year or more.

The warehouse embodies state of the art distribution. Goods-in operations are managed from workstations on the goods-in bays rather than from the goods-in office. Similarly, picking directions are issued to assemblers from "traffic island" stations on the goods-out bays, from where data is downloaded to forearm mounted RF terminals. Intelligent product layout techniques were employed to reduce travel distances within the warehouse. Products are differentiated by their characteristics — such as rate of sale and cube. Efficiency and throughput are then increased through cross-docking and reducing stock movements of the faster lines.

The transport fleet is controlled through a vehicle management system, fully integrated with the order capture and delivery planning software. Store orders are downloaded into the route planner, and from there into handheld terminals in the vehicle cabs — giving drivers full route and job details.

The efficiency of the operation has resulted in the removal of two weeks' inventory from Tesco's supply chain, at the same time as improving availability in the stores it services. Demonstrating real added value through the use of property efficiently.

Toyota

As Britain's fourth largest car exporter, Toyota is a key UK manufacturer. The UK is an important market for Toyota, both in terms of sales and manufacturing. As part of its wider European strategy, the company has established two production centres and a parts distribution centre in the UK. The manufacturing plants are:

- a vehicle plant at Burnaston, near Derby
- an engine factory at Deeside, in North Wales.

Operated by Toyota Manufacturing UK (TMUK), they represent on going investment in excess of £1.7 billion.

Burnaston is the sole production centre for the new Toyota Avensis, launched in the spring of 2003. It also builds three and five door versions of the Toyota Corolla. Deeside produces both petrol and diesel engines for Avensis and Corolla. It also makes engine parts for Toyota's French and Turkish plants and exports to other countries around the world, including South America.

75% of Toyota car production in Burnaston (220,000 vehicles in all) is exported to 16 European countries. Toyota parts distribution centre in Lutterworth holds over 160,000 parts and accessories needed to back the range of products sold in the UK.

In 10 years Burnaston has built more than 1.3 million vehicles, including more than 550,000 Avensis models and more than 350,000 Corollas. Deeside has also enjoyed rapid expansion. In 2000 a new aluminium casting process was initiated and in 2002 production volume was increased to provide machined parts for other Toyota plants as far afield as Venezuela and South Africa.

The Burnaston plant is 230,000 m^2 in size on a 242 hectare site and includes press shop, body weld shop, paint shop, assembly shop, plastics shop, offices, stores and ancillary buildings. A major utilities complex was built to serve the site with electrical power, hot water, chilled water and cooling water for the processes. Around 3,000 people are employed at the site.

The plant has a complete manufacturing operation including press and weld, paint, plastics, engine and assembly plus a comprehensive environmental control facility and an administration building. Parts and logistics and packaging methods have been designed to meet just in time goals. The packaging for components is designed for the use of small returnable packaging throughout the system. To ensure efficient flow of materials, components are collected using a dedicated vehicle fleet. Component suppliers do not deliver direct to the factories.

The logistics system is organised to achieve frequent, regular timed pick-ups of small quantities from each supplier and then a timed delivery is made direct into the plants without any intermediate warehousing.

Many suppliers to Toyota have also been developing their just in time manufacturing capabilities and also aim to hold minimal stocks at their factories. Many are now manufacturing only a few hours ahead of the pick ups and in some cases receive direct electronic broadcasts from Toyotas production lines in order to synchronise their own manufacturing.

Personal and special capacity

Individuals hold interests in land and buildings, other than private domestic dwellings, in various forms; shares in private property companies, partnerships, sole traders and various trusts. The estate manager will advise on the most convenient and tax-efficient form in which the property should be held, having regard to the long term aims of the owners. Individuals are particularly aware of the impact of taxation, due to the directness of its effect, and in order to mitigate tax liabilities, transactions, projects and the receipt of income may be arranged in an economically efficient way so as to obtain the greatest net of tax return. Despite advice, recommendations may not be accepted, due to the owner's expectations, and elements of inertia and uncertainty can offset the opportunities for positive management. Decisions can be delayed until the company or family unit are forced to take action, for example, when business assets are still held by the retired generation of a family business on whose death inheritance tax liabilities are created which could easily have been avoided.

Special legal status can be attached to the management of the property assets of companies or individuals. When a company becomes insolvent then the responsibility for the management of the company or its property assets may be formally removed to parties with that special legal status.

An administrative receiver may be appointed under the Insolvency Act 1986 by a debenture holder who wishes to have control of the company in order to obtain repayment of the outstanding debt. Such a receiver has to be a licensed insolvency practitioner, normally a partner of an accountancy practice, and his powers and duties are strictly regulated by the 1986 Act. Once appointed, though the directors stay in office, their powers are suspended for the period of the receivership and the company may be able to continue to trade.

Clearly, where significant property assets are held by the company in receivership, the receiver will rely heavily upon advice obtained from surveyors in valuing the assets and providing strategic advice on maximising value or effecting disposals.

A debenture or floating charge should be distinguished from a mortgage or fixed legal charge as a mortgagee has alternative powers to appoint a receiver under s 109 of the Law of Property Act 1925. Such a receiver does not have to be a licensed insolvency practitioner and often the mortgagee will appoint a surveyor to this role. The mortgagee's receiver manages the mortgaged property and applies the revenue from it to discharge outgoings and prior liabilities including his own remuneration and then to pay the interest due to the mortgagee, but not to pay off any of the capital loan unless the mortgagee so directs. He is, nevertheless, regarded as the mortgagor's agent and binds him alone unless the mortgage provides otherwise. The mortgagee can in this way avoid the liability which will rest on him if he himself enters into possession of the mortgaged property. Care must be taken, however, if the receiver is not a licenced insolvency practitioner that the appointment is made under the fixed charge only and not any floating charge which may be included within the mortgage documentation. To act under even a limited floating charge without a licence could lead to criminal prosecution by the Department of Trade. While the Law of Property Act 1925 sets down the main roles and duties of the receiver, powers are limited and therefore they are invariably extended by the mortgage documentation.

Both administrative and LPA receivership must be distinguished from the role of a liquidator whose *raison d'étre* is the termination of the company's existence. This may arise under a voluntary winding-up; here a majority of directors must make a statement that the company will be able to pay its debts within 12 months. The shareholders then appoint a liquidator whereupon the directors have no power over the company. Alternatively, the creditors or shareholders can petition the court to make a receiving order. If the court is satisfied that the debtor has no substantial defence, then a receiving order is made and the Official Receiver takes control. This could be followed by:

- a statement of affairs by the debtor
- meeting of creditors
- appointment of a special manager of the company
- public examination of the debtor
- an adjudication order.

After which the debtor's property passes to the trustee in bankruptcy and the debtor incurs the disqualifications of a bankrupt. A trustee may, within 12 months of the appointment, seek leave of the court to disclaim a lease entered into by the bankrupt.

Considerable skill and judgment is required by insolvency practitioners and surveyors involved in this work. There are responsibilities to both the directors of the company, the shareholders and any other parties having a legal interest in the property assets of the company as well as the appointor.

The predominance of mortgage finance for property transactions during the 1980s has led to considerable increase in Law of Property Act 1925 receiverships during the early 1990s, with mortgagees in many instances appointing chartered surveyors rather than insolvency practitioners to manage the mortgaged property and, where appropriate, effect disposal.

The death of individuals results in the operation of the functions of personal representatives whose duties are principally to:

- gather together and realise the assets of the deceased, and make any necessary investments
- pay the debts of the deceased, the administration expenses and the inheritance tax
- distribute the balance amongst the beneficiaries according to the terms of the will or the rules of intestacy.

Personal representatives may be either executors appointed under the will, or in the case of those who died intestate, administrators whose appointment arises when letters of administration have been granted. The powers of management of personal representatives are set out in s 39 of the Administration of Estates Act 1925. As a result surveyors may be instructed to advise upon valuation for inheritance tax purposes and the appropriate method of disposal for property.

Trustees may be appointed in many different circumstances by:

- the settlor
- the beneficiaries
- a person granted that power by deed or under the Trustee Act 1925
- the court.

The surveyor may be involved in acting as a trustee or taking instructions from trustees of several district types:

- personal representatives acting as trustees
- a family trust
- a trust for sale as the vehicle for managing joint ownership
- a charitable trust.

A wide range of statutory powers are given to trustees by the Trustee Act 1925, including under s 23 the right to appoint solicitors, bankers, surveyors and similar agents: and under the Law of Property Act 1925, s 29 which allows trustees for sale of land to delegate their powers to solicitors, accountants, surveyors and other agents.

The investment power of trustees are prescribed as to those securities in which they are permitted to invest under the Trustee Act 1961 and any other securities specifically authorised in the trust deed, as a result the power to invest in real property is limited; though trustees for sale of land are given power under the Law of Property Act 1925 to buy land with proceeds and capital money under the Settled Land Act 1925 may be used to purchase freeholds or leaseholds with more than 60 years unexpired.

Changes in the law governing the rights and duties of trustees and personal representatives were recommended in 1982 in the 23rd report of the Law Commission including all trustees being able to buy freehold or leasehold investments (but not development properties), subject to obtaining appropriate professional advice.

In acting in a special capacity or advising such a client, the surveyor will only be able to provide the best service to a client if he has a clear appreciation of the powers, duties and constraints attached to the legal status. A request to study the formal documents creating the status may initially receive a rather surprised reaction but in almost every case, the client will see the importance of receiving advice in the correct context. It will also demonstrate the need for close liaison between solicitors, accountants and surveyors in adopting a co-ordinated approach to the estate.

Sources and further reading

Company reports — *www.ft.com*

Economic Capital Allocation with Basel II, Chorafas DN, Butterworth Heinemann, 2004

European Association for Investors in Non-listed Real Estate Vehicles — *www.inrev.org*

Finance Acts 2003 and 2004, The Stationery Office Ltd.
Financial Statistics, Government Statistical Service, HMSO, quarterly.
International accounting standards board — *www.iasb.org*
Investment Management Association — *www.investmentuk.org*
Investment Property Databank — *www.ipdindex.co.uk*
Occupier research — *www.occupier.org*
Pension Funds and Their Advisers, Annual.
Promoting more Flexible Investment in Property: a consultation, HM Treasury, 2004
Property law — *www.propertylawuk.net*
The Association of Property Unit Trusts — *www.aput.co.uk*
UK Real Estate Investment Trusts: a discussion paper, HM Treasury, 2005.
Valuation and Investment Appraisal, Darlow, Estates Gazette, 1983.

4 Public Sector Estates

Introduction

Public sector estates grew as the role performed by the public sector developed, up to the early 1990s. Since then they have diminished due to privatisation and disposals arising from more effective asset management. The estates can be distinguished from those of the private sector by a structure of accountability to the public through various officers responsible directly or indirectly to elected members, and by the specialised nature of a substantial part of the estate which meets social and public needs. The predecessors of some of today's local authorities built up substantial urban estates over several hundred years, most notably the Corporation of the City of London.

The public sector estate (both central government and local authority) extends to almost one-fifth of the area of the UK. Asset valuation of properties is carried out by the individual Authorities but due to the area occupied by the Forestry Commission, its value is undoubtedly very much less than that proportion of the surface area. Central government occupies about 8% of the total area, managed by the departments or authorities directly or under contract with the private sector.

Although in 1985, statistics showed a total of 5500 general practice surveyors employed in the public sector this is no longer the case. Since the removal of the Property Services Agency in the early 1990s most Government Departments rely on private sector consultants for specialist advice. The exceptions are where constant design functions are required or valuation work eg Highways Agency, HM Prison

Service, Valuation Office Agency and relevant Departments in Northern Ireland. This does not usually include estate management, most of which is carried out by government officers who are not necessarily qualified surveyors, with the assistance of consultants from the private sector. This is different to the local and metropolitan authorities who have the benefit of professional staff to deal with estate management.

There is some similarity between public sector estates and the estates of commercial and industrial organisations, for each provides a service to enable the occupier/owner to carry out their primary, non-property-orientated activity. Urban land is no longer acquired for public sector purposes but is allocated for designated use particularly in the area of housing development dependant on supply and demand. The development of such is then carried out by the private sector or housing associations who have taken over the previous local authority responsibility for social housing (council housing). Technological developments and the changing strategic situation in the world permits the rationalisation and disposal of surplus MoD lands, which in themselves cover some 1% of the UK .

The estates function within the operational estates of the public sector (and perhaps also commerce and industry) can be summarised as follows:

- productive operational space in suitable premises in the right location to deliver the required service
- the flexibility to meet inevitable change
- acquiring and disposing of property as circumstances require, within the procedures of the estates strategy and approvals system
- using and maintaining the estate in a cost effective manner
- complying with all statutory and legal requirements.

In the past doubts may have been raised as to the vigour with which statutory authorities review their land needs and the speed with which surplus land has been marketed. There are several reasons for this, a natural inertia against disposals, and time-consuming administrative procedures had to be completed. Some elected authorities, as a matter of policy, pursued the acquisition of land within their area. The Conservative Government elected in 1979 saw property not just as a necessary resource for performing the functions of government but as a valuable asset capable of providing funds for the public sector. The ways individual public bodies interpreted and reacted to the policy was influenced by political aspects, the character of the estate and the balance between the merits of freehold and leasehold disposals.

Many forces have been brought to bear on the public sector estate to change these principles. They have been generated by:

- the change in the responsibility of the individual departmental estate
- government policy on privatisation and the sale of assets
- National Audit Office and Audit Commission reviews
- separation of provider and supplies functions (outsourcing of services)
- reorganisation of local government, including the demise of the GLC in the 1980s and the creation of Unitary and Metropolitan Authorities in the 1990s
- asset registers, rents and management plans.

State agencies

Despite numerous privatisations a substantial and varied central government estate almost defies classification:

- Atomic Energy Authority
- British Waterways
- Department for Constitutional Affairs
- English Heritage
- English Nature
- English Partnerships
- HM Revenue and Customs
- Ministry of Defence
- National Rivers Authority
- National Health Service
- Network Rail
- Office of the Deputy Prime Minister
- Office of Government Commerce
- Police and Fire Service
- Post Office
- Regional Development Agencies
- Research Councils
- Scottish Development Agency

- Universities and Further Education Colleges
- Welsh Development Agency.

Increasingly, central government seeks to operate executive agencies within the major departments together with Non Departmental Public Bodies (NDPBs) with clear functions and budgets.

Some groupings can be identified:

- old utilities — waterways, railways, post office
- safety — defence, police, fire, health, with the last of these established as trust status
- development — English, Welsh, Scottish Estate and Regional Development Agencies.

Although the universities and research councils also exhibit characteristics of the private sector and charities.

Government landlord

Property Holdings was formed in 1990 to correspond with the untying of the Property Services Agency (PSA), which was formed in 1972, but its constituent parts had existed for several hundred years. Over the years small discrete central government estates had been amalgamated; the Ministry of Public Buildings and Works was formed in 1963 and this became part of the Department of the Environment on its formation in 1970. Outside advice was taken on how to structure the physical resources of land, buildings, supplies and the supporting technical staff. It was agreed to set up a semi autonomous agency within the DOE with substantial managerial freedom in its day to day affairs, which is rather different from the strong influence of elected members in the management of many local authority estates.

The PSA consisted of those functions already within the DOE concerned with physical resources of central government, together with the Defence Lands Service transferred from the Ministry of Defence. Four basic kinds of services were provided:

- estate management — the provision and management of land and buildings to meet the needs of many different public sector clients
- new works — the provision of services for the design and management of construction works

- maintenance of all kinds of structures and plant
- supplies; the provision of all types of equipment from office furniture to cars.

As the civil estate in the UK alone comprises over 10,000 premises widely scattered and some very extensive in area, the whole operation had to be broken down into a number of manageable units and a decision-making structure. In order to manage the wide range of technical skills, ensure that clients' needs were carefully analysed and responded to effectively, the organisation consisted of headquarters directorates concerned with clients' requirements and the overall management of the staff and resources and operational regions. England formed eight regions, with Wales structured as a region and Scotland with a hybrid organisation. With respect to the maintenance function, each region consisted of several areas and then districts, and it is within the works sections of the regions that the great majority of the staff were to be found. Leslie Chapman's book *Your Disobedient Servant* offers a constructive and critical analysis of the Southern Regional Works Section and has been the cause of much discussion.

The first annual report of the PSA was published in 1978 and this was undoubtedly partly due to pressure from several directions for adequate disclosure. The second annual report contains a brief statement of the PSA's general aim, "to meet its clients' requirements efficiently, economically and in accordance with the Government's policies for the conservation and improvement of the environment. In the light of these policies the Agency aimed to enhance standards of design in new buildings and other works within its responsibility".

The predominant type of property in the Civil Estate is offices, and outside London the majority of these are the local offices of the Department of Works and Pensions which is a combination of the Department of Social Security and Department of Employment. Other mainstream departments have regional offices in the larger town sand cities.

The merits of freeholds are appreciated but public expenditure cuts limit the new development that can be undertaken. In overall terms around 50% of the Civil Estate are freeholds: in London the figure is around 30%. The Hardman Report, accepted by the Government in 1974, aimed to transfer 30,000 civil service posts from London and release 5 million sq ft of leased office space. In 1979 the Conservative Government set in motion a comprehensive review of the civil service manpower; this inevitably included reviewing function, location and resources.

The PSA represented a tremendous management task, with a total staff in 1981–82 of 29,000 serving a very wide range of clients. The professional estate management staff of 270 represent about 14% of all the professional staff and had the support of around 250 technical staff.

Accountability became a general issue and this has shown itself in a number of different ways, parliamentary, as distinct from government, investigations through the Public Accounts and Public Expenditure Committees, and some of the exchanges have been extensively reported, the majority arising initially from comments in Leslie Chapman's book.

In 1982 a businessman, Alfred Montague was appointed Chief Executive. He sought to introduce various measures to encourage efficiency.

- Charging all occupiers for the market rent of their accommodation, in 1984–85 £680m was paid under this system. Reviews of accommodation needs and resources on a town-by-town basis. For example, between 1979 and 1988 it was planned to reduce London Office estate from 21.5 million sq ft to 17.2 million sq ft of which currently only 40% is owner-occupied.
- The use of consultants and contractors which also permits the testing of internal costs.
- Shorter lines of communication and greater accountability for budgets and resources.
- Greater autonomy for the defence estate, perhaps recognised by the appointment of Healey & Baker to advise on the MOD estate in 1984.

The purpose of the 1990 reorganisation was to separate out:

- ownership
- occupation
- property management services.

The result was to form Property Holdings as the landlord of the government estate, to meet the needs of the various occupiers, who increasingly have their own budget for accommodation.

In 1994 the Cabinet Office published a report on the Management of the Central Government Estate. This recommended that Departments should act as principals with intelligent client skills with a much reduced role for Property Holdings in holding information and providing advice.

The recommendation culminated in the authorisation from HM Treasury on 1 April 1996 that Departments should take full responsibility for their estate portfolio with the corresponding accountability. This also signaled the demise of Property Holdings with any relevant information on premises being provided to the individual Departments and with the need to provide estate management advice HM Treasury established the Property Advisers to the Civil Estate (PACE). Within PACE the important advice function was carried out by the Central Advice Unit (CAU). In fact the CAU had been set up in the previous year to prepare for the transition of responsibility and produce guidance documentation. This guidance took the form of substantial binders covering such matters as estate strategy, premises management, appointment of contractors and consultants. These documents were subsequently produced on CD and although they have not been updated for some time they still form an excellent guide to the day to day running of an estate with emphasis on public sector methodology.

There was no longer a "Government Landlord" and the role of PACE also covered the provision of, "off the shelf" advice, clarification of government policy, the organisation of a central database of all the premises on the Civil Estate and consultancy input for which fees were charged. Subsequently re-organisation has lead to the creation of the Office of Government Commerce (OGC), an independent office of HM Treasury into which PACE has been subsumed. The wider role of OGC is to improve the efficiency and effectiveness of central civil government procurement and developing and promoting private sector involvement across the public sector. Major developments in policy followed with the publication of the Lyons review of government accommodation and the proposal for a significant reduction in the size of the civil service with 2.5% annual efficiency savings quantified as £21.5 billion per year by 2008. Part of this efficiency saving was to relocate 20,000 public sector posts out of London and the south east by 2010 with £30 billion of asset sales by 2010.

Alongside this, another efficiency review was carried out by the former head of OGC Sir Peter Gershon which looked at *Releasing Resources to the Front Line* and published in July 2004. This produced robust and detailed proposals which recommended to ministers stretching but realistic departmental efficiency targets for the period 2005/06 to 2007/08 which will deliver gains of over £20 billion.

Within this remit there is still the requirement to consider advice on property and construction issues although latterly it has not been as prominent as when PACE dealt solely with these issues. With regard to the central database this has been developed as an electronic Property Information Mapping Service (e-PIMS) of government's civil property occupations. e-PIMS displays the precise location and outline of department's properties, holdings and occupations on computerised maps and allows users to access and amend their own core property details on-line over the GSI and the Internet. All this is generally confined to the Civil Estate and not the equally extensive Defence Estate which includes all sections of the armed services.

The Defence Lands Service was demerged from the PSA in the 1980s and reconstituted as the Defence Land Agent.

The greater part of the Defence Estate is situated in the UK and one-third is overseas, most of which is in Germany. The management of the Estate has since 1990 been under the guidance of their own internal organisation "Defence Estates" (DE, originally referred to as Defence Works Services). DE have been instrumental in establishing new methods of procuring works and services within the forces primarily to improve efficiency and economies and therefore value for money. They have played a key role in creating integrated project teams (sometimes referred to as prime contracting) in the Building Down Barriers programme which has now become the future policy for procurement and the MoD estate in the UK has been divided into five regional prime contracts which will provide all core works and maintenance services across Great Britain. Apart from meeting the extensive and specialist needs of the forces of the UK, estate management and works are carried out for NATO and for the US Air Force.

At the peak of the 1939/45 war, Service requirements extended to 11,500,000 acres but in a review of Defence Lands in 1947 it was estimated that 1,027,200 acres were required. Between 1961/62 and 1971/72 191,000 acres of land were disposed of and at the latter date negotiations were in progress on 880 individual sales with a total area of 21,500 acres. By 1971 the MOD holdings of land and foreshore had fallen to 662,000 acres and a Defence Lands Committee was appointed to review the holding of land by the Armed Forces to make recommendations as to what changes should be made in these holdings and in improved access for the public, having regard to recreation, amenity or other uses which might be made of the land. This was, therefore, an extensive, rural conservation orientated review, very different from the urban development site based review provided for in the Local Government Planning and Land Act 1980.

This has been assisted by the removal in 1979 of the government's administrative ring fence system whereby surplus defence lands had to be offered to any other part of central or local government before being placed on the market. In 1984, there were 85,900 married quarters, of which the surprisingly high number of 14,500 were vacant. The introduction of the Public Private Partnerships and particularly the Private Finance Initiative has resulted in many projects being funded in this manner and all MoD married quarters were purchased by Arlington Homes and subsequently managed by them.

The peace dividend has resulted in the closure of RAF and naval bases, though with a reduction in the Army of the Rhine, there is considerable pressure to house returning army personnel. With the rationalisation of the estate many bases are now being used for other purposes.

Statutory undertakers

The previous range of statutory bodies, responsible to central government for providing the infrastructure necessary for a highly developed urban society, principally energy and communications have almost entirely been privatised during the past few years. These are commonly referred to as statutory undertakers or public utilities and although in private sector hands, and that includes many overseas organisations, they still have the responsibility to perform in accordance with statutory requirements and are regulated by government bodies. Much of their land is operational land occupied by sophisticated and specialist plant, and some of this may become surplus to requirements and is then sold for other uses in the commercial sector for manufacturing, retail and residential development.

The largest estate in this previous group was that of the Coal Board, with 250,000 acres in the early 1980s, but this has almost entirely disappeared with the onset of privatisation and the introduction of North Sea oil and natural gas. However, the most important in terms of a valuable and complex urban estate was that managed by British Rail, which extends to 191,000 acres.

National Rail (British Rail)

As one of the 10 largest landowners in the country British Rail was perhaps the best example of a complex and specialist estate with considerable development potential. The estate was classified as either operational land, required for use or possible use for railway purposes, or non-operational and available for disposal. The now defunct British Rail Property Board was formed in 1969 and between 1970 and 1981 inclusive contributed, in rental and capital receipts, £435 million to British Railways.

Land within the operational estate was not sold but leased and, if circumstances changed, transferred to the non-operational estate. In 1983 gross income from lettings totalled £71 million. An important area for future income growth was a fresh approach to station trading. The first Smith's bookstall on a station opened in 1848; by 1935 there were over a thousand but since then there has been a steady decline which has now halted. Stations offer extraordinarily high pedestrian flows, many hundreds of thousands of persons per day at London termini. Good layouts in what is called "barrier line trading", occupied by tenants specialising in particular trades, can achieve a very satisfactory weekly turnover. The use of turnover rents results in a joint endeavour by landlord and tenant to boost sales to their mutual advantage.

Land which has been designated as non-operational cannot through its use for railway purposes make a financial contribution to the primary function of the railways but can be sold or let. In 1978 these 30,000 acres were valued on an existing use basis at £182 million. Ownership of non-operational land, together with disused tunnels, bridges, viaducts, embankments and retaining walls, carries with it some onerous expenditures necessary to ensure the safety of the public and the stability of the land of adjoining owners. In addition, nearly 500 buildings are listed as being of historic interest and to some extent the railways have the same problems as the Church as the unwilling custodians of historic but functionally obsolete buildings. The burden has a resulted in being prepared to almost give away historic buildings to local authorities and amenity groups.

Much of the surplus land is the route of old branch lines and where possible sales are made to adjoining owners or local authorities. One of the problems is long leases containing impractical clauses. For example, the railways have some 19th century 999 year leases in which they covenant to return the demise to the freeholder complete with an operating railway system.

For their large development projects based on stations the Board may need a special Act of Parliament, for example, the British Railways (Liverpool Street Station) Act 1983. This enabled Rosehaugh Stanhope to start on a £250 million scheme to rebuild Liverpool Street Station and build 1.25 million sq ft of offices and 30,000 sq ft of shops (Broadgate). The Board received an initial purchase price for each phase, followed by a deferred overage payment. The aggregate of the initial purchase prices was £75 million.

In 1983 the Board completed schemes with a total floor area of 900,000 sq ft and started 25 schemes covering 1.25 million sq ft with £75 million of private sector investment. Perhaps the most impressive scheme completed in 1984 was the 250,000 sq ft Victoria Plaza office building with its three atria.

In pursuing its development programme the Board was faced with highly complex problems of:

- road and rail communication
- services
- listed buildings
- special legislation.

The total rent roll extended to 110,000 items. The vast majority of the rents relating to small charges for rights to cross railway land, where as the heart of the commercial property portfolio is contained in just a few hundred leases. The Board did, of course, make a contribution to the revenue of the railways as a whole and their successful equity participation in future projects is therefore a benefit to every tax-payer.

During 1980 as part of its policy of disposal of public sector assets the Government announced a plan to set up a holding company for some subsidiary activities of British Rail, including shipping, hotels and property. The Board's approach to the detailed management of its portfolio was already very commercial, as developers seeking drainage and other rights over the Board's land in order to release land for development have discovered. The Board take the view, correctly, that this is analogous to the principles elucidated in *Stokes* v *Cambridge* (1961) and seek what they regard as an appropriate proportion of the development value realised as a result of the grant of rights.

In 1985, the Monopolies and Mergers Commission reported on the Board with 43 recommendations for improvements of the "could do better" nature,

In 1992 the future shape of British Rail becomes clear with three principal operating functions:

- intercity
- rail freight
- regional railways

each with their own operational and non-operational estate, making use of the infrastructure. A central track operating organisation, Railtrack, was formed as the start of privatisation with the trains being operated by separate passenger and freight companies. Railtrack, which has now been superceded by Network Rail is an engineering company formed to revitalise Britain's railways. It maintains, improves and upgrades every aspect of the railway infrastructure.

Quasi-commercial estates

A number of social and economic factors have led the public sector into taking some major development initiatives. Arising from the unemployment of the 1930s, industrial estates were created in the assisted areas and post-war planning produced the New Towns, providing a new environment for over one million people. The extent of these programmes at any particular time is influenced as much by political issues affecting policy as technical or economic factors influencing detailed implementation.

The New Towns, must not be seen as solely a public sector concept. The first two New Towns, Letchworth and Welwyn, were private sector activities, and in 1963 Letchworth was successful in getting its assets transferred to what is in effect a trust, Letchworth Garden City Corporation.

Since their inception the New Towns had received capital advances up to 1979 of over £2,500 million and in addition to their social and community services, the total assets include:

- 350,000 houses
- 9 million m^2 of industrial space
- 1.3 million m^2 of office space
- 1,2 million m^2 of shop space.

Each town was developed from a mainly rural setting by an autonomous statutory body. Reductions in national population projections have had their effect on the target populations and development programmes, with the result that during the 1980s more second generation towns are likely to achieve a relatively stable character.

By the early 1960s it was clear that the first generation of New Towns were moving towards a balanced and completed state. As a result the Commission for New Towns was created in 1961 to manage their assets but it was later agreed that housing and neighbourhood shopping would be transferred to the district councils and only the commercial and industrial estates would go to the commission.

The Conservative Government of 1979 reopened the whole issue with a policy decision to sell off £100 million of commercial and industrial assets per annum, and between April 1979 and September 1981 £212 million of disposals were made. There are complex political, social, economic and estate management arguments as to which is the best type of landlord for these assets. Good estate management would favour one landlord able to plan through leasehold covenant and adopt a comprehensive approach to improvement and ultimately renewal and redevelopment. However, the temptation to sell piecemeal to institutions as a way of obtaining capital without selling gilts with an obligation to service the debt proved too great.

The Urban Development Corporations in the Local Government, Planning and Land Act 1980 are the application of New Town methodology to secure the regeneration of urban areas. The first two were in London Docklands and Liverpool, with an initial maximum capital loan of £200 million which can be raised by order to £400 million. This is quite distinct from the borrowing limit of the Commission for New Towns which funds the new towns programme. The Urban Development Corporations are granted very wide powers to acquire, hold, manage, reclaim and dispose of land, to carry out building operations and carry out the functions of statutory undertakers. They will not be operating like New Towns in greenfield locations but in areas with an existing range of local government services. The Act enables the Secretary of State to make a vesting declaration in respect of land owned by local authorities or public bodies in the area, transferring the property to the urban development corporation.

The New Towns and Urban Development Corporations Act 1985 is designed to assist the winding up of new towns and amend the funding of the UDCs.

In response to the depression of the 1930s the government formed North-Eastern Trading Estates in 1936 on a 700 acre site at Gateshead with the long term aim of building 400 factories capable of providing work for 15,000 employees. The Team Valley Trading Estate now employs a workforce of 17,000 and there are still 100 acres undeveloped. In 1960 the English Industrial Estates Corporation was formed, incorporating all three of the Government's industrial estates in the assisted areas.

At the end of March 1984 the portfolio extended to 30 million sq ft with a rental income of £15.6 million and a capital cost of £242 million.

In 1983 84,1,182 units were let with a floor area of 3 million sq ft (approx) but 472 units were vacated by tenants with a floor area of 2 million sq ft (approx).

The corporation does not see itself in competition with the private sector but complementary to the role of the private developer. The emphasis is on manufacturing employment rather than meeting the needs of warehousing and service industry. Further initiatives are being taken to develop inner city sites which the private sector may find unattractive in financial terms but which offer a very real social return.

The Industry Act 1980 permitted the corporation to adopt a more entrepreneurial approach, as a result Legal and General, Barclays Bank and the Coal Board Pension Fund provided £25m of development finance, supported by rental guarantees from the corporation.

The Industrial Development Act 1985 amended the English Industrial Estate Act 1981 by enabling the corporation to, provide advisory services to tenants, obtain grants from the EC and borrow from the general money markets. In the same year the corporation proposed a £350 million scheme to redevelop Chatham Dockyard.

The complexity of the types of quangos involved in land management and development is demonstrated by the arrangement in Wales. The Welsh Development Agency was formed in 1975, taking over the responsibilities of the Welsh Industrial Estates Corporation. It is concerned with three areas of activity; direct investment in industry, environmental improvement and the development of industrial estates. The latter is their main activity and in 1979 they acquired 552 acres and had a land bank of 1000 acres, a further 500 acres being under negotiation.

The majority of its developments are small advance factories, particularly in areas affected by steel closures. The first programme was announced in April 1977, and between then and May 1978, 310 units had been authorised and 52 had been completed. The average size of the units was about 6000 sq ft and some were only 1500 sq ft.

In 1992, the government announced the formation of an Urban Regeneration Agency to absorb English Estates and take over derelict land functions from the DOE. This was subsequently retitled The English Partnership.

Subsequently eight Regional Development Agencies (RDAs) were set up in the English regions as non-departmental public bodies. Their primary role, along with a ninth RDA, the London Development Agency, is as strategic drivers of regional economic development in their region. The RDAs aim to co-ordinate regional economic development and regeneration, enable the regions to improve their relative competitiveness and reduce the imbalance that exists within and between regions.

Responsibility for sponsorship of the RDAs moved from the former DETR to the DTI in 2001.

In addition to the above and totally separate in every respect is the Land Authority for Wales, which has identified itself as filling a vacuum between the allocators of land for development and the developers of land. It believes it has established a clear role for a commercially viable public organisation working within the planning structure but concerned solely with the acquisition, management and disposal of land for industrial, commercial and residential development. In order to clarify what appears to be an overlapping of function, the authority is an entrepreneur in bare land and does not have a budget for construction.

Local government

The growth of local government as the means of providing services to meet the local needs as determined within financial constraints by locally elected members has given a distinctive character to local authority estates.

Local government is big business in terms of the services provided and the resources it controls; education, social services, highways, recreation and in some authorities municipal docks, airports, industrial estates, shopping centres and wholesale markets. Housing now being under the control of housing associations.

With the small districts in rural areas no longer providing housing their estate may include little more than ownership of various office buildings and ancillary accommodation to perform their statutory duties. Whereas the large urban districts comprising cities of several hundred thousand population, such as Sheffield, Leicester and Bristol are each the largest landowners within their urban areas, involved in commercial property management and entrepreneurial projects.

The Bains Report (1972) identified the importance of the estate management surveyors in the allocation of resources in local government:

> Land and buildings are among the most expensive and scarce resources of a local authority and the efficient management of these resources is a matter of first importance. In many authorities the management of existing land

> and buildings will be a substantial task, which, coupled with the acquisition and disposal programme, may well justify the existence of a separate major department closely linked to the land utilisation sub-committee of the policy and resources committee. We have in mind particularly authorities with a substantial investment in land and buildings and also authorities with large town/city centre redevelopment programmes in hand.
>
> Each authority should create and maintain a central record of all estate owned, and should co-ordinate all forward land acquisition and disposal proposals in order to maximise the use of existing estate. In particular consideration should be given, wherever possible, to the combination of separate acquisition and building proposals on to one site, in order to achieve savings in building costs and in common services and for the convenience of the public.
>
> Through their land utilisation sub-committee authorities should also consider carefully and keep under review the economic use of existing property. Changing local circumstances may suggest that it would be more economic to sell or redevelop existing sites and rebuild on a different site where the demand can equally well be met. The possibility of combining commercial and public uses should not be overlooked.

Local government is not one comprehensive structure. During the second half of the 1980s the Conservative Government abolished the GLC, creating unitary authority London boroughs. Much was lost in the political expediency. The professional approach adopted by the GLC to staff development has a long tradition, as this extract from the *London County Council Staff Gazette* of 1900 indicates:

> The municipal officer's education is not completed when he has left school or college, and as some of the profoundest economical and social questions are being worked out in the problems which every day arrest the attention of municipal government it is evident that officers, to be abreast of the work required of them, must be students as well as officers ... and a professional officer is doubly important if besides being a specialist in his own profession, he knows some of the influences which have produced his work and which in turn his work is likely to produce.

In 1974 England and Wales were divided into metropolitan and non-metropolitan counties and London boroughs for the purposes of local government. These metropolitan and non-metropolitan counties in turn consisted of metropolitan and non-metropolitan districts. The 1990s local government reorganisation introduced unitary and non-unitary authorities in England and Wales while Scotland was divided into 32 council areas.

England

Apart from the agricultural smallholdings in the shire counties the largest specialist statutory estates in local government are social services and education, other than where the Private Finance Initiative has been introduced to operate schools. Their importance is reflected in their budgets and staff, which dominate their respective authorities. The estates departments service their needs as clients in a rather similar way to the in-house estates departments of commerce and industry. On the other hand, there is a property management and an entrepreneurial role where the estates department takes a lead, earning an income for the council and influencing the visual character and economic structure of the area, often the most important means of achieving positive planning.

The estates surveyor works within overall policies created by a complex framework of individual elected members, committees and the corporate management of senior officers. The following job description of the chief officer of a city estates department illustrates these points and the wide range of professional duties, these are listed below.

Professional responsibilities

Advises the council, committees, chief officers and heads of departments on all aspects of land and property interests, including mortgages and loans, prepares the relevant reports and valuation and represents the council, when necessary, at tribunals and public inquiries.

Provides professional support and expertise to the city council's economic development policy and, in particular, gives high priority support to the industrial development officer, including practical assistance where necessary.

Undertakes the negotiations by way of purchase, lease or other arrangements for all land and property required by the council, and claims for compensation related to property, including work for "agency" services.

Negotiates the disposal by way of sale, lease or other arrangements for all land and property surplus to the council's requirements.

Undertakes the efficient management and related maintenance of the council's corporate estate, central shopping area, commercial and industrial estates, the markets and other land and property held for general or redevelopment purposes; with particular emphasis and sensitivity to new industrial developments and nursery units, especially where they support the development of small firms.

Undertakes the maintenance of such public buildings and the management and/or maintenance of such other buildings, land and property as may from time to time be the responsibility of the department.

Operates a continuous review of the council's overall land holding, their land and property requirements. As part of the corporate activity, identifies and promotes the redevelopment potential of the council's land interests in conformity with their economic development and planning strategies.

Maintains a complete record of the council's interests in land and property, maintains and updates the relevant Ordnance Survey map records and other documents, keeps adequate records and files of the department's activities and proper books of account, and undertakes the "referencing" of private ownerships.

Maintains and updates valuations of the council's properties for fire insurance purposes, to accord with the city treasurer's requirements.

Prepares estimates of revenue and expenditure in connection with the department's activities for the information of the council, committees and management team.

Maintains an effective working relationship and joint approach with other public authorities and in particular with the private sector, to assist in the efficient operation of the council's functions and the attainment of the city's aims, objectives and economic prosperity.

Departmental management responsibilities

Controls the activities of the department and expenditure of the departmental budget, to meet approved targets, programmes and services.

Reviews the performance of departmental staff at regular intervals and advises on appropriate training.

Manages and organises the staff of the department in order to meet defined objectives.

Delegates appropriate responsibility to senior staff, consistent with effective decision making, while retaining overall accountability.

Subject to the policies of the appropriate committee, selects and appoints departmental staff.

Ensures compliance with statutory obligations and council policy governing the safety, health and welfare of departmental staff and their visitors.

Undertakes such other duties consistent with the appointment as may be required by the appropriate committee or the chief executive.

Corporate management responsibilities

Participates fully in the corporate management process as a member of the chief officers' management team, including the provision of professional advice and expertise as well as contributing to the activities and business of the team generally.

Discharges the duties and responsibilities of the post with due regard to the principles of corporate management and the policies of the council.

In counties and districts with a substantial urban base, implementation of these responsibilities can require between 20 and 50 qualified estate management staff, with appropriate technical support. Much would depend on the extent of the investment estate and the involvement in maintenance. Some idea of the scale of the estate can be seen in a typical county with 3,000 holdings and a total fire insurance reinstatement value of £1,000m. During a year there were 485 acquisitions and 106 applications for planning approval related to the corporate estate. The departments of the authority together occupied about 250,000 sq ft of offices. One of the organisational problems ie whether to structure estates departments on a geographical basis; allowing staff to identify closely with their area, or on a functional basis related to the type of work, enabling the staff to further develop their professional expertise. In addition it is important that the main programme committees are attended by experienced officers knowledgeable in the detail of estates matters which influence the provision of services. The effect of these factors can result in a multi-tier structure requiring a high level of personal as well as professional skill in its senior members.

The introduction of Compulsory Competitive Tendering (CCT) from 1995, when authorities were required to outsource up to 65% of their construction related services largely converted the Authorities into clients, with only a limited contractor role. While CCT is no longer carried out in its previous form the result has remained the same and contracts are required to be competitively procured.

In small districts, particularly in rural areas, estate matters tend to be the responsibility of the chief executive or planning department, with consultants advising from time to time. Authorities may call in the district valuer, who provides a consultancy service. This is often suitable for specific valuation advance, but that office has limited experience of continuing estate management work.

The incorporation of further education colleges, with effect from April 1993, removed them from county council control, as was the case of the polytechnics during the 1980s. In a less comprehensive way the same is happening to secondary schools; much more change should be expected, including the impact of city academies. See Chapter 8, occupiers, for current practice.

Positive management

The management of local authority estates is influenced by the particular holdings and geographical boundaries of the authority and the detailed management must have regard to statutory requirements such as s 123 of the Local Government Act, although many of the former constraints upon disposals of land and property by local authorities have been removed.

As central government seeks to reduce or control by capping the proportion of its contribution to local government finance the entrepreneurial estate of the authority can take on an important role. First, by its management in the interests of the Council Tax payers, additional sources of revenue and capital are essential in order to maintain the existing provision of services without unacceptable increases in the Council Tax.

In order to benefit from the urban infrastructure that they provide authorities need a land bank which can be brought forward for development either by the council itself or by a developer on a ground lease. Whether land should be sold to realise the maximum capital value or leased at current open market rent is a matter of investment policy and the view of the elected members as to their responsibilities to current rate-payers and future rate-payers, with the benefit of the impartial advice of the estates surveyor.

Second, the estates department has a major contribution to make towards positive planning. It can assist the planning department in advising on cost and value implications of its policies and the likely reaction of the private sector to various possible planning briefs. As the largest land owner in the area, the estates department can make plans reality by implementation on local authority sites, which may encourage the private sector. This will require a relatively subtle working relationship between the two departments, in which conclusions can be reached on any conflicts between social planning objectives and entrepreneurial estates objectives. This area of activity is one in which the planning and development surveyor is making a major contribution in seeking to bridge the gap between planners and developers.

Third, the estates department is uniquely placed to take initiatives in the fields of industrial and economic development. This can be done by assisting and consolidating the existing industry and the encouragement of natural growth by the provision of nursery units, though the enthusiasm of the planners for high-quality layouts and landscaping will have to be curbed if the pioneering industrialists are to be able to afford the rents. In fact, it may be better to create nursery units by refurbishment of existing buildings in the older industrial quarter of town than by provision on peripheral industrial estates. The attraction of new industries is an altogether more complex task. It requires the corporate marketing of the character of the town in the best possible light together with an imaginative and attractive response to any leads.

In emphasising the commercial role of the estates department it must be recognised that this can only be carried out within a local government framework. The estates department must respect the political constitution of the authority and the importance of implementing the statutory functions but in its general management of the commercial estate it should adopt an approach to the market close to that of a large and responsible property company or institution. It is not always easy to separate the political, the commercial and economic factors.

The Audit Commission has stressed the need for authorities to adopt policies which lead to rational decisions about the advantages and disadvantages of ownership. This means property with suitable; tenure, location, arrangement and condition for the effective, efficient and economic delivery of services.

This requires a management system which recognises three levels of decision-making:

- Political — clear committee responsibilities and increasingly a lead member
- Technical — property expertise in a department with a strategic overview and relevant technical skills
- Service department/user — encouraged to see property as a valuable asset, responsible for day to day management.

Local government finance can cause distortion between capital and revenue items, affecting all three levels.

Sources and further reading

Annual Reports, Welsh Development Agency.

Devolving decision-making: refining targets and performance management, Cabinet Office, HMSO, 2004.

Lyons Review, HM Treasury, 2004.

Management of the Central Government Estate, Cabinet Office Efficiency Unit, May 1994.
Releasing resources for the frontline, Sir Peter Gorston, July 2004
Whose property is it anyway? RICS, 2002.

See also *www.hm-treasury.gov.uk* for latest and past government reports.

Charities

Historical development and the Charity Commissioners

From the early Middle Ages the wealthy and worthy have endowed various bodies with assets for specified charitable purposes. The motives of these benefactors varied, in some cases it may have eased their consciences, others may have seen it as easing their way to heaven, and others made their gifts in the very best sense, out of charity. Over the years and centuries the assets rose or fell in value and were changed, the specified purposes of the charity became more or less relevant, and the whole institutional and social environment appeared quite different to succeeding generations of trustees.

The importance attached by society to charities is shown by several hundred years of specific statutory provisions and, perhaps most important from the point of view of the management of the charitable funds, exemption from taxation. Widespread malpractices by trustees in the 19th century led to the establishment of the Charity Commissioners in 1853, and their present supervisory function is governed by the Charities Act 1992.

The draft Charities Bill published in 2004 included a wider range of activities which would qualify as charitable purposes, but removed the presumption of public benefit from educational charities.

The Charity Commission have adopted a much more rigorous approach to accounts and the accounts of the 300 largest charities are available on their website.

Charities are involved in real property in three distinct ways.

- The largest charities have been endowed with substantial property investments and land, once agricultural, is now the site of many commercial developments.
- The charities in carrying through their purposes need to occupy both general and specialist property. This is an operational use similar to that of any occupier of property, but much of the property will be owner-occupied, arising from gifts of many years ago.
- A small number of charities are in the specific business of the preservation of land and buildings, characterised by the National Trust.

The history of charities is an integral part of our social history as the core from which the whole of the Welfare State has developed. The founding and endowment of the Oxbridge colleges in the 14th century and hospitals, as early as 1215 in the case of St Thomas', has resulted in their ownership of substantial agricultural and commercial property holdings. To these can be added the City Livery Companies and Merchants' Guilds of the older provincial cities. The largest charitable estate is that of the Church Commissioners, and there remains the Crown Estate, almost defying classification, but in many respects exhibiting management characteristics similar to those of the largest charities. All the above have some special features but they possess sufficient common features, such as their antiquity and close involvement with the establishment, to be called collectively the old or traditional institutions.

It is important to distinguish charities from private trusts; the former benefit the public at large or a section of the community, can exist in perpetuity and do not fail if their objectives become frustrated. The Charities Act of 1960 replaced many previous overlapping public and private Acts of Parliament and the Charity Commissioners now operate under its provisions in promoting the effective use of charitable resources by encouraging the development of better methods of administration, by giving trustees information, advice and checking abuses. About 190,000 charities are registered, but a small number are exempt, including most of the old institutions, specially the church and the Oxbridge colleges.

Significant amendments were made by the Charities Act 1992, ss 32 to 36 set out criteria trustees must meet in order to grant short leases or make any other disposal of an interest in land without obtaining Charity Commissioners approval, though the procedures are similar to those of the Charity Commissioners.

Under s 23 of the 1960 Act the Commissioners can authorise a charity to acquire land with the benefit of a surveyor's report on valuation and other relevant matters. Further, larger charities can be given a general authority to purchase investment property. When selling land charities are under a duty to show they have acted scrupulously and obtained property advice and used this to obtain the best price. The Commissioners approval must be obtained unless this is accepted by the 1992 Act or vested in the Official Custodian of Charities by the High Court or Charity Commissioners.

From the point of view of the character and management of their property investments the larger landowning charities, the old institutions, can be distinguished from the new institutional investors in a number of ways.

They have been in existence for several centuries and the most important, such as the church, have acquired their own statutory setting for the management of both their investment and operational land or are subject to supervision by the Charity Commissioners.

Property has been added to the investment portfolio in a rather random way in the past by bequests and there has tended to be a presumption against sale. Until recently, portfolios have not been planned or managed with quite the vigour adopted by the new institutions.

The relatively modest inflow of new funds, unlike pension funds, means that in order to carry out development it may be necessary to sell some of the portfolio to realise sufficient finance, or alternatively, and more likely, the charity will grant a ground lease, participating in the development value by rent reviews.

There is overall, a cautious investment policy, with less opportunity for large profits and losses, further the limited disclosure requirements means that transactions and investment performance are not often reported in the technical press. The cautious investment approach is reflected in small or negligible debts, hence little or no gearing, resulting in greater stability than that of property companies. There are specific allocated purposes for the revenue from investments which can influence the investment strategy.

Charities, as part of their *raison d'être*, imply a commitment to some patent standards of morality, and this can influence the type of property/tenants which the trustees will find acceptable. Further, tenants may seek to make negotiating factors out of this aspect.

Exemption from tax, apart from VAT, is almost absolute. This has implications on the attractiveness of particular investments which carry onerous tax implications for typical investors.

They have a higher proportion of their assets in property than the new financial institutions and particularly in agricultural land, which has proved to be well placed as greenfield development sites.

The estimated total value of assets of charities is over £75 billion. The largest 500 charities account for nearly 50% of the total income of £32 billion.

It is possible for trustees to vest property in the Official Custodian, who holds 3,500 parcels with a value of £1 billion. With over one million charity trustees and evermore onerous regulation replacement of trustees (as for school governors) is of increasing concern to incumbents.

The Church Commissioners

For hundreds of years the estate of the church was held by individual cathedrals, chapters, dioceses, bishops, parishes and other ecclesiastical church units. During the 19th century the Ecclesiastical Commissioners acquired more of these assets and by 1891 they had an income of over £1 million. The present structure of the Church Com-missioners was created by the Church Commissioners Measure of 1947 and the major role of the investment funds was to meet clergy stipends. Their expenditure on contributions to stipends, pensions and clergy houses account for approximately 90% of their investment income. The investment policy is to obtain the best current income consistent with maintaining the real value of the assets, but they do bring ethical and social as well as financial criteria to bear on their investment policies and practices. Property has been making a growing contribution to the income of the Commissioners. In 1970, Stock Exchange investments provided £12.6 million and property £8.4 million (net).

The commercial property has its origins in historical episcopal and chapter estates now occupying some key sites in London, and during the 1960s and 1970s a considerable amount of development work was carried out. Recently, the emphasis has changed to portfolio management and rearrangement of leases.

In 1956 the Commissioners formed Church Estate Development and Improvements Company as a holding company and it has been very active as a means of carrying our projects jointly with other companies. In 1991 they owned shares in 17 companies in England and Wales and 25 in the USA.

In 1991 with over half their assets in property, often large scaled risky developments, investments funds suffered, a programme of sales was implemental and an enquiry was set up by the Archbishop of Canterbury into the performance of the Commissioners.

Since then development has largely ceased and the property is held in large investments. The 20 largest urban and rural properties accounting for 87% and 73% respectively of each portfolio.

Currently, of the £4 billion of fixed assets, just over £1bn is held in property; with a value of:

commercial	£522 million
rural	£288 million
residential	£334 million

The commercial portfolio has a 30% weighting in the West End, the total return of 7.7%, under performed the IPD benchmark of 10.7%, in 2003.

The substantial agricultural estate showed a return of 21.9%. The residential property, most of which is in central London, is largely split between the Hyde Park Estate (1989 units) and the Octavia Hill Estate (1590 units) with a 10.6% return.

The fate of redundant churches causes considerable concern; between 1969 and 2003, 1656 churches were declared redundant with the following consequences for uses:

Alternative uses	
private chapels	26
arts	37
storage	35
museums	40
misc	42
commercial	58
worship, Christian	120
monuments	139
residential	213
civic	238
	948
Demolition and site disposal	
housing associations	69
local authorities	48
church yards	47
new places of worship	63
community	24
other	115
	366

Preservation		
Churches conservation trust	332	
Department of Environment	4	
Diocesan Board of Finance	6	
	342	
		1656

Source: Church Commissioners Annual Reports

Of the £33 million proceeds raised since 1969, £6.4 million has been allocated to the Churches Conservation Trust, which is jointly funded with the Department for Culture, Media and Sport; the latter contributes at a rate of £3 million pa.

The Crown

The Domesday Inquest of 1086 showed that the Crown owned one-fifth of the entire kingdom. The present Crown Estate has evolved as a result of complex statutory arrangements and possesses a unique status which almost defies classification. The Crown Estate must be distinguished from the private Crown Estate; the former is the public property of the crown, held in "right of the crown", the latter the property owned privately by the Queen, including Sandringham and Balmoral. In 1760 the crown surrendered the Crown Estate in return for a civil list payment. Since 1810 it has been administered by a board of commissioners and the present Crown Estate Commissioners hold office under the Crown Estate Act 1961.

There is no organisation in the world quite like the Crown Estate. With a property portfolio encompassing many of the UK's cityscapes, ancient forests, farms, parkland, coastline and communities, the Crown Estate's role as guardian, facilitator, manager, influencer, employer and revenue creator is unique.

> We have two main objectives: to benefit the taxpayer by paying the revenue from our assets directly to the Exchequer and to enhance the value of the estate and the income it generates.
>
> Our portfolio has a value in excess of £4 billion, demanding the utmost in professionalism and efficiency from our staff and management to create added value for everyone — the people, businesses and communities who deal with us, and the nation as a whole.

> The estate performs no statutory function and the duty of the Commissioners is to maintain the Crown Estate as an estate in land (with such cash or investments as may be required for the discharge of their functions) and to maintain and enhance its value and the return obtained from it, but with due regard to the requirements of good management.
>
> *Crown Estate Annual Report* 2004

The estate is not in receipt of other sources of income nor is it subject to taxation and the emphasis is therefore on portfolio management rather than acquisitions.

The estate is one of the most diverse and comprises six elements.

The urban estate is distinguished by highly reversionary interests and its concentration in London in Regent Street, Regents Park and Kensington. Outside London, their focus on quality schemes in cathedral/county towns has resulted in less exposure to risk than many other developers.

Housing is committed to social schemes; handicapped, those on low incomes in key local services and historic buildings.

The agricultural estate extends to 270,000 acres.

The marine estate poses diverse economic political and public issues; coastal zone management, marine aggregate extraction, moorings and wind farm licences.

Windsor Park and Forest owned for 1000 years attracts two million visitors a year.

The Scottish estate is a microcosm of the other estates plus fish farming with a key role in employment in rural areas.

The 2004 report shows:

	Capital value £ m	*Rental £ m*
Offices	1614	103.3
Retail	816	45.6
Rural	508	14.9
Minerals	85	16.4
Marine	216	18.1
Residential	632	168
Other (urban)	509	16.2
Fish farming	11	1.5
Forestry	18	0.1

Source: Crown Estate Annual reports

The core values of the estate are, commercialism, integrity and stewardship. The estate aims to seek a long term balance in its management between outperforming the Investment Property Databank Index (in 2004, net total return of 13.9% against the benchmark of 12.4%) and qualitative values. The latter is delivered through:

- a customer management team
- corporate and social responsibility
- environment management policy
- disability access
- investors in people
- marine stewardship.

There are eight commissioners and just under 400 staff split equally between administration and operational.

Other investment estates

The colleges, hospitals, livery companies and municipal charities can take a longer term view than property companies but lack the financial resources of the new institutions. By definition most of their assets are old having been bequeathed as farmland, possibly developed in the 19th century, redeveloped in the 1920s and subject to feasibility studies for a further redevelopment. The portfolios are often not so large as to be capable of balance in the sense of the large institutional funds and each fund thus acquires its own character. In 1978 Massey and Catelano estimated that five charities or quasi-charities held something over 2 million acres, in the following percentages of the total area of the UK as agricultural investments.

The Crown Estate	0.48%
The monarchy	0.30%
The Church Commissioners	0.28%
Oxbridge colleges	0.34%
Other charities	2.60%

To see this in context, this can be compared with the work of the Northfield Committee (1978) and Steel (1983) on the holdings of the financial institutions. Northfield estimated the institutions held 530,000 acres, Steel, four years later, estimated this had risen to 880,000 acres worth £1,148m, held as follows:

Pension funds	£415 million
Insurance companies	£585 million
Property Unit Trusts	£148 million

The Oxbridge colleges hold in total about 200,000 acres of agricultural land within their total property portfolio. Trinity College, Cambridge, is probably one of the wealthier colleges, in 1971 its total portfolio of investments was worth £23 million, producing an income of £1.5 million. The property sector of the portfolio consisted of 255 houses, 76 shops, 15 warehouses and 16,000 acres of agricultural land.

The municipal charities arising from schemes of amalgamation are among the largest land owners in a number of towns and are closely involved in redevelopment proposals as ground landlords. There is an Association of Land Owning Charities, in which the Oxbridge colleges are prominent.

Operational estates

The property of charities which is used for charitable purposes rather than investment purposes is usually owner occupied. This is rather similar in principle to the property of manufacturing industry where the estate manager is mainly concerned with maintenance and repair. If the premises were endowed and built at a time when labour was cheap, say in the 1920s, such buildings now require heavy maintenance programmes as components require renewal. Furthermore, the overall design is wasteful of space and probably showing signs of technical obsolescence.

Where income from the endowment is insufficient to maintain the operational property, then the trustees will need to enter discussions with the Charity Commissioners to determine how their own affairs can be rearranged — changes which are being made to other charities could be used to their advantage. The Oxbridge colleges, though in receipt of substantial investment incomes, are also faced, like the Church, with the upkeep of historic buildings. The largest colleges each require several hundred thousand pounds per year from their investment income to maintain their mediaeval courts and chapels.

A small number of charities, most notably the National Trust, have as their prime purpose the preservation of historic buildings. The National Trust is very sensitive to the fact that it is funded from private sources, but with the status of a charity and the attraction to donors of exemption from capital gains tax and inheritance tax it is in a relatively privileged position.

The National Trust with over three million members, founded in 1895 is one of the largest conservation bodies in Europe with control of over 600,000 acres and 600 miles of coastline.

Its core purpose is "Looking after special places forever for everyone". The long term strategic aims are:

- showing leadership in the regeneration of the countryside
- deepening understanding of our cultural heritage
- putting life long learning at the heart of everything we do.

Nationally there were 13 million paying visitors to houses and 50 million visitors to coast and countryside.

The most prominent region is Devon and Cornwall. In Cornwall alone the Trust owns 25,000 acres including the majority, 161 miles, of coastline.

In the region, 30 properties are open to the public with 200,000 visitors to each of St Michael's Mount and Lanhydrock, there are also 20 separate car parks and 50 let cottages. A typical large house may have 50 staff, more than when the house was lived in by its old donor family.

Hence, a charity formed to preserve the heritage of land and buildings is also a major tourist business employing in the region over 1000 staff and several thousand volunteers, with retail, catering and holiday letting income of the order of £10 million pa, (this ignores admissions and subscriptions).

The Landmark Trust performs a similar role in respect of domestic property and somewhat exceptionally manages Lundy Island on lease from the Trust.

Heritage property offers the unique opportunity to study the public, private and voluntary (charity) sectors performing very similar tasks.

The public sector is represented by English Heritage whose objectives are:

- improving understanding of the past by research and study
- promoting the historic environment by opening up our properties and increasing access through education
- protecting our historic places and ensuring change is managed sensitively.

English Heritage manages over 400 properties including; World heritage sites, monuments, castles, houses, abbeys, forts, stone circles, the vast majority much older than the National Trust Estate, which explains its *raison d'être*, there are over 5m visitors to staffed sites of which Stonehenge and the Tower of London are the two most defining and visited heritage sites in the country.

In addition, there is a grant making role, £39m pa and a statutory role in advising on 15,000 applications for planning, listed building and ancient monument consents. The income of English Heritage at £154 million was just under half that of the National Trust, of which three quarters came from central government, unlike the Trust.

Several hundred private stately homes, opened to keep the tax man at bay are now large businesses; Chatsworth, Beaulieu, Warwick Castle, Castle Howard, Longleat, to name but a few.

The motivation, objectives and mode of operation of each of the three sectors (public, private and voluntary) of heritage property vary and it is usually apparent from quite small features which type of management is supervising the property. Also each operates under a different financial regime and a somewhat random method of property allocation, influenced as much by the availability of funds as by a ranking of architectural or historical features. Pricing policy is influenced by both general factors and the particular features of individual properties. Nevertheless, the end result in all three sectors is the preservation of heritage property and its enjoyment by in excess of 100 million visitors a year. Also, probably a greater variety of property and effectiveness than if any one of these sectors were given overall supervisory powers to the exclusion of the other two.

Since 1976 the aggregate number of visitors to each sector has steadily risen but in the private sector they have been divided between a growing number of competing properties, with the result that some individual properties have suffered a fall in numbers. The vagaries of the weather on certain key weekends also introduces a distortion into the figures which can mask more fundamental changes in visitor behaviour.

As a final thought, the 40 or so cathedrals of England including Canterbury, Durham, Salisbury, St Paul's and Westminster Abbey while in the care of the Church of England a charity, are increasingly exhibited as if in the private sector and need public sector support; a rich heritage.

Sources and further reading

Annual Reports, Charity Commissioners.
Annual Reports, Church Commissioners.
Annual Reports, Crown Estate.
Annual Reports, National Trust.
Capital and Land, Massey and Catalano, Arnold 1978.
Financial Institutions, Their Investments and their Agricultural Land Ownership, Steele, Reading University 1983.
Funding Recreational Development, Stapleton, 251 EG 841.
Report of the Committee of Inquiry into the Acquisition and Occupancy of Agricultural Land, HMSO, 1979.
Whose Land is it Anyway? Norton Taylor, Turnstone, 1982.

Part 2

Lease

Commercial Property

6

Introduction

The activities and procedures dealt with in this chapter represent what is traditionally the major part of the work of estate managers. It is not proposed to approach the subject from either a legal or valuation standpoint, but one which endeavours to identify decisions the estate manager needs to take in relation to individual leases, and the factors to be considered. Some of the main matters to be considered (from the points of view of both the landlord and the tenant) in arriving at a strategy are set out in Appendix B. As in previous editions of this book, only a limited amount of case law has been mentioned, because, as well as referring to the many legal texts, it is now possible to search a number of databases of case law on the Internet. One such subscription service is the Westlaw UK Landlord & Tenant online resource, which includes *Woodfall: Landlord & Tenant*, and the *Landlord & Tenant Review*. It offers an authoritative source of case reports and legislation but with the additional advantage of being updated daily and providing extensive search functions. Another extremely useful source of case law and articles is to be found in EGi from Estates Gazette. The valuations required by estate managers are not complex but are best approached within the discipline of a clearly developed conceptual base rather than as ad hoc aspects of estate management.

The primary legislation, that is to say that legislation concerned primarily with the ordering of the relationship between landlord and tenant of business property, is found in the Landlord and Tenant Acts of 1927 and 1954, the latter as amended by the Law of Property Act

1969, the Landlord and Tenant (Covenants) Act 1995 and, most recently, by the Regulatory Reform (Business Tenancies) (England and Wales) Order 2003.

In the post-war period, and particularly since the start of the 1970s, there has been a plethora of secondary legislation and administrative procedures that have significantly affected the position of occupiers and owners of commercial property. It is intended to summarise all the primary legislation, but as regards the secondary legislation, its existence will be identified but not pursued.

The last half of the chapter concerns itself with the interpretation and implementation of covenants, emphasising those aspects where particular estate management skills are required. It must be stressed that a comprehensive and detailed legal treatment of such important covenants as those of repair or rent review can only be obtained from a specialist legal text and the necessary advice from the legal profession.

Landlord and Tenant Acts

Two principal Acts and some limited amendments create the statutory framework. The Landlord and Tenant Act 1927, Part I, and the Landlord and Tenant Act 1954, Part Ill, deal with compensation for tenant's improvements at the end of the lease. Part II of the 1954 Act sets out a detailed code of rights and duties at the termination of tenancies, and a number of amendments were made by the Law of Property Act 1969, following a report by the Law Commission. After much discussion over a number of years, the Regulatory Reform (Business Tenancies) (England and Wales) Order 2004 finally updated from June 2004 the previous legislation by making the renewal or termination of business tenancies quicker, easier, cheaper and, hopefully, fairer. The Landlord and Tenant (Covenants) Act 1995 was introduced to enable tenants to limit their future liabilities under leases which they have assigned but it is not retrospective and so only relates to assignments of leases granted after 1 January 1995.

Under the common law, any tenancy that is for a fixed period terminates by effluxion of time at the end of the period. A periodic tenancy terminates on expiry of a notice to quit given by the landlord to the tenant or vice versa. In the case of a business tenancy, the tenant may have built up a substantial business while trading at the premises that are the subject of the tenancy and could lose all the goodwill if he has to leave and is unable to find suitable alternative premises. At the same time, the landlord should be able to remove an unsatisfactory tenant or to refuse to renew his tenancy.

The aims of these pieces of legislation are therefore:

- to provide some security to the tenant by providing for automatic continuance of the tenancy after its normal termination date on the same terms as before until compliance with certain notices and procedures that may terminate the tenancy
- to allow the tenant to require his landlord to grant him a new tenancy
- to limit the circumstances in which the landlord can refuse to renew the tenancy
- to give the court power in certain circumstances to order the grant of a new lease and to settle the terms of it
- to allow the tenant, in the event of the tenancy not being renewed, to claim compensation in some circumstances
- to ensure that commercial landlord and tenant relationships work well, that the parties are treated fairly and that there is encouragement to maintain and enhance the physical fabric.

As was said by Goulding J in 1979:

> Parliament did not by this legislation intend to petrify the economy of business premises. I think rather that the intention was to leave the market to develop freely subject only to the amending protective policy of the Act.

In fact, it is understood that only a relatively small proportion of applications made under the Act result in the court making an order as to the terms of a new tenancy or refusing to grant a new tenancy.

It should be noted that an increasing number of landlords and tenants are now using PACT (Professional Arbitration on Court Terms). This uses a third party solicitor or surveyor acting as an arbitrator or an expert to determine the terms for unopposed renewals without the risk of the court deciding terms that may be less favourable to the parties.

The application of the legislation to business tenants under s 23 of the Landlord and Tenant Act 1954 has been interpreted broadly to the benefit of tenants in the occupation of a holding carrying out a wide range of activities. The extent of this illustrated by *Groveside Properties Ltd* v *Westminster Medical School*, 1983, where a flat leased by the school was occupied by four medical students. The court accepted that a major medical school was a business activity and that part of this business was the fostering of a collegiate spirit, which the flat assisted.

It should be noted that a business activity does not have to be trading mainly for the purpose of making a profit. This was demonstrated in the case of *Hawkesbrook Leisure* v *The Reece Jones Partnership*, 2003. Hawkesbrook Leisure claimed that their solicitors had negligently failed to make an application to the county court for a new tenancy within the statutory time-limits. The solicitor's defence included the claim that Hawkesbrook Leisure were not actually carrying on a business and so were outside of the protection of the 1954 Act.

Hawkesbrook Leisure was a company, limited by guarantee that had been set up to take over sports grounds and social clubs formerly owned by London Transport making them available to members of the general public as well as London Transport employees. Although it was allowed to make a profit, its memorandum of association would not allow it to distribute any such profit to members as dividends or bonuses; any profits were to be used to maintain and improve their leased sports ground facilities.

Since the definition of "business" in s 23 (2) of the 1954 Act includes "... a trade, profession or employment and includes any activity carried on by a body of persons whether corporate or unincorporated", the High Court found in favour of the defendants. Thus, the tenants were carrying on a business within the definition in the Act.

Certain tenancies have been excluded by s 43, including those not exceeding six months unless they contain ... provision for extension beyond six months or the tenant has been in occupation for a total period exceeding 12 months. Also, s 38 provided that on a joint application to the court by the two parties, the court may approve an agreement excluding the provisions of ss 24–28. This procedure, known as "contracting out of the Act" is now used widely and the 2004 reforms removed the need to obtain prior court approval (which was little more than a "rubber-stamping exercise"). For any new tenancy granted since 1 June 2004, it only requires the landlord to serve a prescribed notice on the tenant at least 14 days before the parties enter into the agreement. The 14 day period can be waived in cases of urgency but the tenant must sign a declaration confirming that he has received and accepted the consequences of the notice.

Some of the problems of comprehension and interpretation of the statutory provisions, particularly where there is a consequentiality of events, can be overcome by the application of decision-making techniques. Various bodies such as the Building Economics Bureau and PACE (now OGC) have produced useful flowcharts through the legislation, though (as with any application of such techniques to an existing situation) it does require a certain flexibility of approach in its practical application. A simplified flow-chart is set out in Appendix C.

An essential requirement for the operation of statutory provisions of this type is motivation by the parties to initiate and respond to notices within the prescribed timetables (although the timetables have been made less restrictive by the 2004 legislation). Towards the end of a lease, a mesne landlord between a freehold or long leasehold and a tenant in occupation is under little incentive to operate the system. The Act provides (in s 44 and schedule 6) for the concept of a competent landlord, being the landlord with the most immediate interest for a term greater than 14 months at the date of the relevant notice. There can only ever be one competent landlord and only he can operate the statutory procedures for terminating tenancies.

Security of tenure

As mentioned previously, the Landlord and Tenant Act 1954 provides security of tenure for the tenant. The Act sets out how tenancies may be terminated and the procedures for doing so.

When laying the draft of the Regulatory Reform Order before Parliament, the Minister, Tom McNulty said:

> We do not plan to change the basic framework of the Landlord and Tenant Act 1954. But its workings do need an overhaul. Following extensive consultation, our proposals will modernise and streamline the procedures, for the benefits of tenants and landlords alike.
>
> The proposals include new safeguards for tenants agreeing to exclude security of tenure ("contract out") while removing the need to obtain court permission. The tenant will have to receive a "health warning" explaining in plain English the implications of contracting out, normally at least 14 days before signing up for a lease.
>
> They will remove the need for unnecessary litigation and will help to promote better working relationships between owners and occupiers of commercial property.
>
> "Business tenants will face less risk of missing deadlines and losing the right to renew, following the removal of some notorious time-limit traps in the present law.

Thus, the Regulatory Reform Order has not changed the principles, primarily it has changed the procedures with regard to:

- a relaxation of the strict timetables for notice and counter-notice procedures
- critical dates for issuing proceedings, the nature of their issue and who can issue them
- interim rent applications
- the issue of what comprises a group of companies
- contracting out procedures, as discussed above.

S 24 still provides that no periodic tenancy or tenancy for a fixed term exceeding six months of a holding to which the Act applies shall come to an end unless terminated in accordance with the Act. The only ways that the tenancy can be terminated are:

- by one of the common law methods, notice to quit by the tenant or surrender or forfeiture
- by one of the special forms of statutory notice:
  - tenant's notice to terminate under s 27
  - landlord's notice to terminate under s 25
  - tenant's request for a new tenancy under s 26
- by the landlord and tenant agreeing to the grant of a new tenancy, whereby the current tenancy terminates on the date of commencement of the new tenancy.

The action to bring to an end a tenancy to which Part II of the Act applies can be taken by either landlord or tenant. A landlord's notice under s 25 must be served within the period of six months to 12 months prior to the specified date of termination. The landlord's notice must also state whether the landlord would oppose the tenant's application to the court for a new tenancy and if so the grounds upon which the landlord would oppose the application, or, if he is willing to grant a new lease, the landlord's proposals as to term, rent and other provisions. It must also include a "health warning" that the tenant has a statutory right to a new lease and that the landlord's proposed terms can be challenged through the lease renewal process. The previous requirement for the tenant to serve a counter-notice within two months to protect his statutory rights has been removed.

A tenant for a term of years certain exceeding one year who has not received a s 25 notice from his landlord may request a new tenancy by serving a notice under s 26 in the period six months to 12 months before the expiry of the original tenancy, and in some circumstances it can be advantageous for the tenant to take the initiative in this way.

A landlord who delays service of his s 25 notice may be pre-empted by the tenants' s 26 notice, requesting a new tenancy and specifying a renewal date, 12 months hence, despite the contractual term expiring sooner. In this way the tenant gains up to six months at the existing historic rent.

If the tenant has served a s 26 notice, the landlord cannot serve a s 25 notice, but may within two months serve a counter-notice on the tenant specifying the grounds upon which he will oppose the tenant's application either as to the renewal itself or the proposed terms.

The form of the various notices are to be found in the Landlord and Tenant Act 1954 Part II (Notices) Regulations SI 1957 No 1 157. SI 1967 No 1831 and a quite extensive consolidation in SI 1983 No 133, the latter extending to 70 pages; these are further consolidated following the 2004 Reforms. In order to be sure that the notice is correctly worded and served it best left as a matter for solicitors, particularly as in *Robert Baxendale Ltd* v *Davstone*, 1982, the Court of Appeal reversed the decision of the County Court in permitting an extension of the time limit under county court Rules Order 13, rule 5.

The interpretation of the various "x" months statutory notices was considered in *Dodds* v *Walker*, 1981. This confirmed the general rule that where a notice is given in months, the period of the notice ends on the day of the month which bears the same number as that on which the notice is given (ie a notice given on the third of the month expires on a third of the month). Where the initiating event date month is higher than the last day of the terminating month, then that last day is the terminating date. Hence, a response on 31 January to a notice served on 30 September was out of time.

During the lengthy negotiations for renewal circumstances may change, or the tenant may become aware of the strength of the landlord's grounds for opposition. As a result, the tenant may wish to withdraw his application. In the county court the tenant can withdraw under CCR Order 18, rule 1, without leave of the court. In the High Court the tenant can withdraw under RSC Order 21, rule 3 and as amended by SI 1982 No. 1786, this discontinuance may be without leave of the court.

Notices under ss 25 and 26, while initiated by the different parties and commencing with different procedures, are but two mutually exclusive routes to the same destination; the right of the tenant to make an application, under s 29, to the court within two to four months of the original notice for a new tenancy. The success of the legislation can be judged by the very small number of cases that result in a hearing, and the system has been accepted by professional advisers as generally fair and equitable. The parties, aware of a secured access to the court are generally able to reach agreement. If the landlord has taken no action and the tenant wishes to terminate his fixed term or continuing tenancy then he can serve notice in accordance with s 27.

Landlord's opposition to application for a new tenancy

There are seven grounds specified in s 30 upon which the landlord may successfully oppose the tenant's request for a new tenancy. Some are based upon the tenant's failure to perform covenants, others on the landlord's intentions. In the latter case the tenant may successfully oppose the landlord's intentions through s 31(a) or qualify for compensation on quitting in accordance with s 37, as no blame can be attached to him.

Surveyors will be called upon to give expert evidence for the parties in a hearing before the court and much of that evidence will be related to one or more of the grounds of s 30(1).

a. Breach of repairing obligation. The landlord will have to show that this is of a serious nature and the court may accept an undertaking by the tenant to remedy the breach.
b. Persistent delay in payment of rent.
c. Other serious breach of covenant or any other reason connected with the tenant's use or management of the holding. This grants substantial discretion to the court.
d. The landlord is willing and able to provide suitable alternative accommodation on reasonable terms to meet the tenant's requirements, including the protection of goodwill.
e. Possession is required of the subtenancy in order to let the premises as a whole which are currently let in parts and could be let more advantageously as a whole. This ground will only be relevant when the interest of the tenant's immediate landlord is shortly to end. Only then will the superior landlord have the status of a "competent" landlord.
f. The landlord intends to demolish or reconstruct or carry out substantial works of construction and he could not reasonably do so without obtaining possession. In *Betty's Cafe Ltd* v *Phillips Furnishing Stores Ltd*, 1958, it was shown that the intention must

be established at the time of the hearing, and this can be shown by the stage to which contract documentation has progressed.

The work of demolition or reconstruction must involve either the whole of the premises or a substantial part. In *Morazzi* v *Global Grange Ltd*, 2002, the landlords proposed works costing £2 million over a 12 month period to an hotel were held not to qualify because they did not affect a substantial part of the premises. It was held that the issue was the true character of the works rather than the time or money.

Works of construction that have been held to qualify include the:

- installation of a new staircase, larger lifts and new toilets, concreting of an open area
- installation of tie rods.

Prior to 1969 the landlord could be successful even if he only required part of the premises or required them for a limited period of time. S 31(A) now provides that if the tenant is prepared under the new lease to give the landlord access and/or take a new tenancy of an economically separable part of the holding, then the tenant can defeat the landlord's application. By implication, a landlord could design his scheme so that it causes maximum interference for the maximum period of time and so defeat the tenant's counter-claim.

The work entailed for both landlord and tenant in responding to or initiating matters related to s 31A(1)(a) Access in order to carry out works; s 31A(1)(b) New tenancy of economically separable parts, is likely to be both extensive and expensive, so that in all but the most valuable property this is an encouragement to negotiation.

The terms of the existing lease to be renewed need careful study, if as in *Heath* v *Drown*, 1973, the lease contains clauses giving the landlord wide powers to enter and carry out works, he will have to prove his proposals cannot be accommodated within these provisions.

g. The landlord intends to occupy the premises either for business or residential purposes, but the landlord must have held the superior interest for at least five years. In *Cam Gears* v *Cunnigham*, 1981, the tenant occupied a car park in connection with their occupation of nearby premises. The landlord wished to occupy the holding by erecting a building and facilities for MOT testing of cars; this was accepted by the court.

The case of *Chez Gerald* v *Greene*, 1983, where the landlords claimed they wished to occupy the premises to run a restaurant, illustrated the typical issues where the court has to consider whether the landlords ideas have; (to quote the elegant phrasing of Asquith LJ), "Moved out of the zone of contemplation — out of the sphere of the tentative, the provisional and the exploratory — into the valley of decision.î This can be demonstrated by a resolution in the minutes of the company, the raising of finance, the acquisition of specialist trading expertise, the financial appraisal of the project, consideration of planning and other requirements.

An interesting situation can arise when the lessee is also the planning authority as in *Westminster City Council* v *British Waterways Board*, 1983. Westminster was using the property as a cleansing depot and indicated it would not give planning permission to the board who wished to use it for Waterways purposes. The court held that planning permission probably would be granted on appeal, and that the board genuinely intended to carry out its proposals; the House of Lords confirmed the board was entitled to occupation. If the court is not satisfied on grounds e or g, but would have been satisfied if the date for termination had been up to a year later, it must make a declaration to that effect without an order for a new tenancy.

Subsection (3), which was added by the Law of Property Act 1969 extends the definition of landlord's proposed occupation to include a company controlled by him. This led to the interesting case of *Ambrose* v *Kaye*, 2002 where the renewal went to court on ground g but the landlord gave no evidence in court as to his shareholding in it. During the lunch adjournment his wife transferred the majority of the shares in the company to her husband giving him the controlling interest in the company. Despite protestations from the tenant, the court held, on appeal, that the tenant should have identified this issue at an earlier stage and the landlord could have then dealt with the transfer earlier. It is established law that the court must deal with affairs as they stand at the close of proceedings under the 1954 Act, ie in this case the landlord was granted possession for occupation by his company.

If the court sees the proposals as more distant, it may agree to the new lease containing a break clause or being shorter than would normally be granted, as in *Wig Creations* v *Colour Film Services*, 1969. In *Adams* v *Green*. 1978, the court noted that if a break clause was a disadvantage to the tenant, his rent should be reduced to reflect this. The court was particularly helpful to the landlord in *Amika Motors Ltd* v *Colebrook*

Holdings Ltd, 1982, where in recognition of the landlord's intention to develop, the court granted a five year term with a break after three years.

Interim rents

From the date the contractual tenancy ends to the date the terms of a new tenancy are determined by the court or finally agreed between the two parties can be a long period. Tenants were able to delay final agreement for many months or even years and during this time only the old rent could be recovered by the landlord, while the landlord could himself use such a ploy for over-rented properties to continue receiving a rent at above market rental value. To deal with the former, the Law of Property Act 1969 introduced a s 24(a) enabling the landlord to make an application to the court for an interim rent after he has given notice under s 25 to terminate the tenancy or the tenant has requested a new tenancy under s 26. The reforms brought about by the Regulatory Reform Order extended this to enable the tenant to make such an application and made it payable from the earliest date at which a s 25 or s 26 notice could have been served (ie expiry of the tenancy), whereas for such terminations prior to 1 June 2004, the interim rent became payable from the date of the landlord's application under s 24(a) or the date specified in the landlord's or the tenant's notice, whichever is the later.

The full new rent will also be the interim rent if a new tenancy is granted for the whole of the premises let to the tenant, the tenant is in occupation of the whole and the landlord did not oppose the renewal. The court can, however, be asked to vary the interim rent where either:

(a) the amount would have been substantially different had it reflected market conditions at the start of the interim period or

(b) where the terms of the new lease differ materially from those of the old lease and the amount of the new rent would have been substantially different had the terms not been changed.

The interim rent will usually be fixed at the hearing which also determines the new rent. The interim rent is to be backdated to the ascertained interim commencement date.

Where the above conditions did not apply (eg the tenant was not in occupation of the whole or the landlord opposed renewal) the basis of valuation of the interim rent is on an assumed annual tenancy. This valuation is retrospective, whereas the new rent can take account of things likely to happen up to the date when the new term will commence. If the arguments are accepted that tenants should pay a premium on market value for reviews of periods longer than five years, then an annual tenancy is worth rather less than the general level of market evidence. This proposition does not necessarily apply in a bear market where generally tenants prefer shorter terms. The terms of the new tenancy may be different from those of the old tenancy; the interim rent is to be fixed in terms of the old tenancy whereas the new rent reflects the terms of the new tenancy. Lastly, and most confusing of all, the interim rent must have regard to the old rent. In *Fawke* v *Viscount Chelsea*, 1979, the Court of Appeal accepted that this implied a cushioning effect, though in that case they reflected the failure by the landlord to carry out repairing covenants which resulted in serious dry-rot damage.

Where there are no problems of breach of covenant, then the decisions of the courts have tended to give a discount for interim rent of 10–30% of that amount determined as the new rent. The scale of discretion which the section gives to the judge as valuer is considerable, therefore the role of the expert witness is crucial. For example, in *Charles Follet Ltd* v *Cabtell Investments Co Ltd*, 1986, the landlord suffered a "cushion effect" of 50%. The interim rent element of the evidence cannot be dismissed as less important than that for the new rent. It may only apply for a few months, but the opportunity it provides for demonstrating the application of sound valuation experience and the ways it can be distinguished from the new rent may enable the court to assess the quality of the expert evidence and so strengthen or weaken the effect of the expert's evidence on the court's decision on both rents.

The new tenancy

Arising from the earlier procedures, the court may under s 29 make an order for a new tenancy. This raises a number of issues which, if not agreed upon between the landlord and tenant, present the court with the task of determining all the terms of a new tenancy, the most important of which is the rent.

The Act provides in s 34 that it shall be the rent that the holding might reasonably be expected to be let at in the open market by a willing lessor, disregarding:

- any effect on rental of the tenant's previous occupation; any goodwill attached to the premises
- any effect on rent of improvements referred to below
- any value attributable to a licence to sell intoxicating liquor if it appears to the court that the licence belongs to the tenant.

Improvements are to be disregarded if carried out other than as an obligation of the lease and either carried out during the tenancy about to terminate or completed not more than 21 years before the application for a new tenancy, and the tenant and user have qualified under the Act during that period of time.

The issue of how valuers should give effect to this apparently simple legal instruction on improvements has been considered in *GREA Real Property Investments Ltd* v *Williams*, 1979. The tenant's valuer argued that the premises should be valued as if the tenant's works did not exist and this should be done by deducting from the annual value of the complete premises the annual equivalent of the current cost of the tenant's improvements. The landlord's valuer argued that the same ratio of value for unimproved and improved values that existed at the start of the lease should subsist throughout the whole period of the tenant's occupation. Though this was a consultative case stated from a rent review arbitration, Forbe Js, considered this in its broadest context and provided the following guidelines:

- the tenant is to be credited with the rental equivalent of the improvements
- the improvements are to be assumed to have been paid for when they were done and assessed as a wasting asset
- the improvements should be valued as existing as at the review (renewal) date and not completed at that date.

Later that year he heard *Estates Projects* v *Greenwich London Borough*, 1979, which concerned four former houses used as shops on the ground floor and offices on the upper floors. The court was not enthusiastic about the methods adopted by landlord, tenant or arbitrator but indicated qualified support for the relative proportion of values at the start of the lease, which had been used by the landlord in the previous case.

Where the improvements are a replacement of some previous buildings, structures or site works, rather than an addition, then the landlord will wish to enjoy the hypothetical value which those demolished premises would have had if they still existed. The importance of detailed records including photographs cannot be overstressed.

It has also been suggested that the landlord should enjoy the rental value (if any) attributable to the potential of the unimproved property for its suitability for making improvements.

Care should be taken to avoid the problem that arose in *Euston Centre Properties* v *H & I Wilson*, 1983, where "the tenant" under an agreement for a lease carried out works. It was subsequently held that during the agreement for lease the "tenant" only had the status of licensee and hence could not enjoy the statutory improvements disregard.

The court is given a discretion in s 33 to grant a tenancy for a term not exceeding 15 years (it was 14 years prior to the 2004 changes). This, together with the provisions for the effect of improvements upon rentals, has some interesting implications for negotiations between landlord and tenant.

Consider the following circumstances, bearing in mind that s 34(2)(a) and the 1954 Act provides that the 21 year disregard runs from the date of statutory application for the new tenancy (up to 12 months before the end of the tenancy).

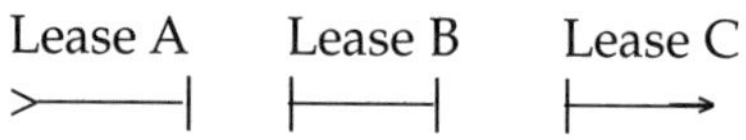

Lease A   Lease B   Lease C

If improvements are completed six years before the end of lease A and if the tenant can obtain in lease B a term of 15 years or less, then there will be statutory improvements disregard at the start of both lease B and lease C. Hence, there is an incentive for the tenant to pursue the exercise of his rights before the court (who cannot grant a term for more than 15 years). By contrast the landlord should (out of court) grant lease B for as long a period as possible, in excess of 15 years; in order that the 21 year disregard is avoided at the start of lease C, enabling the landlord to reap the benefit of the improvements.

Alternatively, if improvements are completed nine years before the end of lease A the landlord wants lease B to be for 14 years and so lease C can commence outside the 21 year disregard. Whereas the tenant wants either a term of 13 years for lease B, so that lease C commences within the 21 year statutory disregard, or as long as possible lease term B, to delay the start of lease C.

There is ample scope here for some significant tactical negotiation in respect of the rental value of different length leases, quite unrelated to the frequency of rent reviews. In *CBS UK Ltd* v *London Scottish*

Properties Ltd, 1985, the landlord was seeking a 14 year term, but the tenant only required a lease for two and a half years, which was granted by the court.

In default of agreement between the two parties s 35 enables the court to determine any other matters and they shall have regard to the terms of the current tenancy and to all relevant circumstances.

This leaves considerable discretion to the court, as was demonstrated in *Cairnplace Ltd* v *CBL (Property Investments) Co Ltd*, 1983. The current tenancy had a provision whereby on assignment the landlords could require two directors of the assignee company to guarantee performance of the obligations under the lease for the remainder of the term.

The landlords sought a provision in the new lease requiring the tenants to provide sureties. The tenants challenged this on two grounds: first, that sureties should be dealt with by s 41A(b), and, second, that the jurisdiction did not extend to third parties. The Court of Appeal confirmed the landlord's request saying, "there may be many other circumstances, differing widely from those in the instant case, in which it would be fair and responsible for the court to determine that there should be guarantors of the tenant's obligations in the new lease".

Assuming that one of these relevant circumstances is the terms being agreed in the market at the date of the new tenancy, then landlords may endeavour to make significant changes to the terms of the old lease.

In *Card Shops Ltd* v *Davies*, 1971, the court refused on renewal of the tenancy to replace a qualified covenant against assignment with an absolute restriction since it would significantly restrict the tenant's future ability to realise goodwill on an assignment.

In *Charles Clements (London) Ltd* v *Rank City Wall*, 1978, there was a restriction of use to cutlers (a trade in which the tenants had been engaged for many years), which the landlord wanted to relax. It was agreed that the difference in rental value was £1750 pa, and the court held that it was not right to impose the relaxation on an unwilling tenant.

Institutions prefer a "clear" lease, that is to say one where either the tenant is responsible for all outgoings, or the landlord, whilst responsible for outgoings, is able to recoup their full cost in a comprehensive service charge. If a landlord can alter the terms of a lease on renewal of a tenancy this can reduce the yield by in excess of 1% and significantly increase the value of his interest, for only a relatively modest reduction in the rental to reflect the more onerous obligations. In the case of *O'May* v *City of London Real Property Co Ltd*,

1979, the landlord was initially successful in introducing a comprehensive service charge in the new lease, with provision for a sinking fund for capital items. The court adopted four tests which had to be satisfied before amending the terms of the old lease.

- has the party demanding the change shown a good reason?
- will the party resisting the change be adequately compensated?
- will the adjustment materially impair the conduct of the tenant's business?
- is the variation fair and reasonable between the parties?

When the case was subsequently heard in the Court of Appeal in 1980, although the tests were approved, the court did not accept that the second test was satisfied. They did not feel that a 5% reduction in rent passing was adequate compensation for a short-term tenant taking on an indeterminate risk that should properly be borne by the freeholder or a long leaseholder, this was confirmed by the House of Lords.

Though these criteria were used for a change in the repairing obligations, they are equally applicable to changes in other terms of the lease and demonstrate the flexibility built into the legislation, and one of the reasons for its success.

One of the most helpful determinations under the Act can be found in *Newey & Eyre Ltd* v *J Curtis & Son Ltd*, 1984, this considered in relation to warehouse premises the similarities and differences of comparables:

- access
- floor area
- rent-review period
- sitting tenant v vacant possession
- repair covenants
- relativities of values between floors
- age and layout
- user clause
- improvements
- premiums and upwards only reviews.

The court fixed a rent of £13,000, which analysed out of £2.40 sq ft for ground floor warehousing, the lease provided for landlord's repair and restriction on user to the storage of electrical goods.

In most instances landlord and tenants reach agreement without recourse to the courts and a tenant agreeing to these types of changes in lease-terms will ensure that the rental is adjusted accordingly. Further, the effect of the lower capitalisation rate creates a situation similar to marriage value, and the tenant may well seek a share of the value added to the landlord's interest. This will be strongly resisted by landlords, but is a useful negotiating point for tenants' surveyors.

The Act also provides in s 32 for the court to determine the property comprised under the new tenancy. In almost every case this will be unchanged but where the tenant is prepared to accept part only and the court rejects s 31A, (1)(b) landlord's grounds of opposition, the tenancy shall be that smaller part.

If the court orders that a new tenancy be granted, the tenant has 14 days within which to apply for the order to be revoked, presumably in case he changes his mind on learning what the terms of the new tenancy are going to be.

Compensation on quitting

If the landlord is successful in satisfying the court upon one of the seven grounds in s 30, then s 31 provides that the Court shall not grant a new tenancy. However, while some of the grounds are based on the tenant's breach of covenant, others are based solely on the landlord's initiatives and no blame can attach to the tenant. A landlord who is successful under s 30, paras e, f and g, will, in accordance with s 37 (as amended by s 149 and schedule 7 of the Local Government and Housing Act 1989), have to pay compensation, on a basis which possesses only the merit of certainty. The compensation is twice the rateable value if the tenant has been in occupation for more than 14 years and once the rateable value for any shorter period. Since rateable value approximates to a year's market rental value on FRI terms , the successful use of paras e, f and g can be very expensive for the landlord. The Secretary of State can vary the multiplier. Where the landlord patently has a good case it would be a total waste of the tenant's time to apply for a new tenancy, and the section as amended for a new tenancy in order to protect his right to compensation.

A problem arose in *Edicron Ltd* v *William Whitely Ltd*, 1983, in defining the holding upon which the rateable value multiplier should be based. Here the tenant reoccupied a portion of the premises that had previously been sublet between the service of the s 25 notice and after putting in an application for a new tenancy but before the expiry of the s 25 notice. The tenant subsequently discontinued his application for a new tenancy and under section 64 the tenancy expired three months thereafter. The court held that the date of the service of the s 25 or s 26 notice defined the extent of the holding, as a result compensation at four and a half times the rateable value was obtained on a part which had not been occupied for 14 years. Whereas in *Card Shops Ltd* v *John Lewis Properties Ltd*, 1982, the date for determining the multiplier that applies is the date the tenant quit.

In the event of properties which have only been partly occupied by the tenant, the compensation will be based upon the rateable value of that part occupied. In these circumstances the Valuation Officer can be asked to provide an apportioned valuation.

Finally, tenants who receive the statutory compensation will be pleased to know that it is not charged to Capital Gains Tax. The Court of Appeal in *Drummond* v *Brown*, 1984, held that the payment was not "a capital sum derived from an asset ... in particular a capital sum received by way of compensation for any kind of damage or injury to assets or for the loss destruction or dissipation of assets" as provided in the Capital Gains Tax Act 1979. Rather, it was simply a sum which Parliament said should be paid, the decision is perhaps generous to the taxpayer.

Compensation for improvements

If tenants are to improve their business premises then various incentives are required, and these include adequate compensation for the tenant quitting on the termination of the tenancy. The tenant will only qualify under Part I of the Landlord and Tenant Act, 1927 if, prior to the commencement of the works, he serves notice on the landlord, who has three months in which to object. If there is no objection and the landlord does not offer to do the works himself then the tenant can carry out the works. If the landlord objects then the matter can be determined by the court. The court must be satisfied that any improvement is of such a nature as to add to the letting value of the holding at the termination of the tenancy, is reasonable and suitable, and does not diminish the value of any other property belonging to the landlord.

It is thought, somewhat surprisingly, that *Deerfield Travel Services Ltd* v *Leathersellers Company*, 1982, was the first case to give guidance on what constitutes sufficient notice by the tenant, under s 3 of the Act. The court held that a series of documents served by the tenants did constitute proper notice and that the landlord's responses did not constitute an objection within three months. Hence, the court could certify the works carried out were proper improvements. The compensation shall not exceed the smaller of the net addition of the value of the holding as a whole which is the direct result of the improvement, or the reasonable cost of carrying out the improvement less the cost, if any, of putting the improvements into a reasonable state of repair.

The effect of the improvements on the value of the premises is influenced by the landlord's subsequent intentions. If these include demolition, structural alterations, refurbishment or change of use, then these will be reflected in the basis of compensation, to the detriment of the tenant. In practice, the 1927 Act procedures are seldom initiated by tenants. In shop property many so-called improvements are of a trade rather than a property character and most leases provide that at the end of the term the tenant shall restore the premises to their original condition.

In practical terms, any compensation payable under the 1927 Act procedures is most often offset against the landlord's claim for outstanding dilapidations at the end of the lease.

Request for information by either party

The s 40 notice is a notice requesting information, a mechanism by which the tenant can or the landlord can ascertain exactly who is the appropriate landlord or tenant with statutory protection. From the reforms coming into effect in 2004 the procedure has been tightened up. Whereas the party upon whom the notice had been served could ignore it if they wished, they now have to take it seriously and the court can order a recalcitrant party to respond..

Either party may serve a s 40 notice during the last two years of the tenancy requesting details as to the name and registered address of the other party who has to provide a response within one month otherwise action can be taken as a breach of statutory duty. As it is not restricted to one such request, a property owner may serve another such notice just prior to serving a s 25 notice to check that it is still the same occupier (for example, there may have been a sub-letting).

Tenant vacating

If the tenant does not wish to renew the lease and is proposing to vacate the premises at the expiry of the lease, there are two options.

1. Vacate the premises at the expiry of the contract; it is not necessary to hand the keys back to the landlord or his agent, although it does provide further evidence of the parties acceptance of the situation. This has been held by the Court of Appeal to take the lease outside of the 1954 Act. The leading cases are *Esselte AB* v *Pearl Assurance plc*, 1997 and *Single Horse Properties Ltd* v *Surrey County Council*, 2002. The Regulatory Reform Order has in its changes to s27(1) taken into account the Esselte decision on the grounds that the landlord should be fully aware of the expiry date and can inspect accordingly at that time.

   However, despite being given statutory effect by s 27(1A) the method may be considered to be not totally without its risks. If the tenant has not fully vacated the property but has left any items of furniture for example, or has not fully complied with all covenants relating to dilapidations, the court may take the view that he has not properly terminated the tenancy.
2. Serve a s 27(1) notice, which need not expire on a quarter day but on any day more than three months from the notice being given.

Landlord and Tenant (Covenants) Act 1995

This Act was introduced to address the perceived inequitability of some well publicised instances of original tenants being pursued by landlords for rent arrears of subsequent tenants but without their having any rights of re-occupation of the property. Often in these cases the original tenant had long since retired and assigned his lease which had then been further assigned several times over the length of the lease. Under the terms of the original contract (the lease) the original tenant had privity of contract with the landlord and so remained liable for all subsequent liabilities.

The Act only applies to leases which commenced from 1 January 1996 onwards and provides that the original tenant is only liable for the first assignee (who should therefore be known to him), but once the assignee assigns to another party, the original tenant ceases to be liable; his assignee ceases to be liable on a subsequent assignment and so on. In order to provide protection for the landlord the Act provides that the assignees should covenant directly with the landlord and introduces the concept of an AGA (Authorised Guarantor Agreement) by which they covenant to indemnify the landlord against claims against themselves when they are either tenants or their first assignee.

Secondary legislation

The common law relationship between landlord and tenant is set out in the lease, which should be a comprehensive contract between the parties. In fact, the lease may not cover every eventuality and it also has superimposed upon it the Landlord and Tenant Acts. Both the lease and the Landlord and Tenant Acts are concerned with the legal rights, powers and duties between the two parties directly concerned with the use and occupation of the demised premises.

Social, technical and economic change and the ineffectiveness of written agreements have shown the need for a much wider pattern of rights and duties between individuals, and between individuals and the state. In relation to property, the law of tort, particularly nuisance, waste, negligence and trespass, represent a substantial pattern of rights and duties in respect of third parties. Some wrongful acts may constitute both a breach of contract and a tort, and in such circumstances the plaintiff may choose the basis for his action. In recent years statute law has tended to take over the role previously played by the law of tort in protecting third parties and seeking to ensure that the public at large are not subject to various dangerous conditions or undesirable practices. These provisions represent a substantial volume of secondary legislation and are either of a general environmental type or directed at a specified group of persons; owner, occupier, owner-occupier or the employer. Unless there is some indication to the contrary, they tend to be directed at whoever is in possession of the premises.

General environmental measures

The pattern of public health legislation and planning legislation, together with some housing legislation, grants to local authorities substantial overall control of the environment. In some fields it is a technical matter of measurement of quality, in others a qualitative opinion of safety, health or visual amenity. Within the breadth of the environment law some of the provisions are concerned with the creation and maintenance of the infrastructure of services which permit the existence of intensively developed urban areas. Others refer to the pattern of rights and duties between the community as a whole, represented by the local authority, and the owner and occupiers of individual properties. In a few cases the infringement of this pattern of rights and duties may constitute a public nuisance and the local authority may be able to exercise its powers under the Public Health Act 1961 to deal with dangerous structures and carry out works if the owner fails to take the necessary action.

The depth and complexity of the provisions can be seen from such sources as within the Sweet & Maxwell Local Government Law Encyclopaedias. Here, as long ago as 1994, Environmental Health, in three volumes, and Planning, in four volumes, each contained over 60 statutes dealing with the quality of the environment and the physical condition and use of land and buildings. Since then there has been a considerable proliferation of further legislation, much of it stemming from the EU.

The manager of urban estates will from time to time have to deal with the effects of this type of statutory provision exercised by local authorities or various government agencies with which he has had no previous contact, and despite many years of experience, may be unaware of their detailed content. It is invaluable to have access to and the ability to use the various reference and research systems which are increasingly becoming available. An attitude of mind favourable to the interpretation of new statutory and administrative material is an essential requirement for those now entering the profession as is a continual updating on the new legislation and interpretation thereof.

The general nature can be shown by the Health and Safety at Work Act 1974 which is an enabling act, under which regulations such as the Control of Substances Hazardous to Health Regulations 1988, which include a reference to *legionella* bacteria. In April 1991 the Health and Safety Commission approved a code of practice for The Prevention or Control of Legionellois. This requires the risk to be assessed and minimised, staff trained and records kept. Risk management is an increasing area of work for the property manager. The 2005 criminal case against a council official in Barrow in Furness who failed to take precautions against legionella bacteria demonstrates the seriousness for the property manager.

The Environmental Protection Act 1990 contains extensive new powers and duties for local authorities.

The most significant was the duty under s 143 to set up a Register of Land which may be subject to contamination. The Environment Act 1995 retrospectively inserted powers for an enforcing authority (a local authority or the Environment Agency) to serve a remediation notice on the owner or occupier of contaminated land, closed landfills or special sites. In the first place remediation is the responsibility of the person or persons who caused or knowingly permitted the land to be contaminated. However, if after reasonable enquiry, no such appropriate person can be found, the owner or occupier bears the responsibility for undertaking the remediation. It should not be assumed that this only applies to major contamination by hazardous waste as in some cases local authorities have threatened action on owners of land where litter and waste have been allowed to be blown off their land onto adjoining council land.

One of the first cases to appeal under the Environment Act 1995 is that of *Circular Facilities (London)* v *Sevenoaks District Council* 2004. A developer was held liable for remediation costs even though his predecessor caused the pollution and the developer no longer had an interest in the land, having developed it for housing and then sold the homes. The developer was held liable as being responsible for introducing householders as "receptors" of the pollution.

Another important case with far-reaching implications for property owners, managers and developers is that of *Van Der Walle* v *Texaco Belgium*, 2004 — a European Court of Justice case. Petrol leaked from a service station storage tank; the soil and groundwater contaminated by this were held to be "controlled waste" under the Waste Framework Directive (and enacted in the Environmental Protection Act 1990). Allowing it to occur or keeping it on site without a Waste Management Licence constituted an offence. Furthermore, if it were considered to be used as landfill, it would require Pollution Prevention Control (PPC) permits as a deposit or keeping of waste materials. This case broadens the effect of the Environmental Protection Act to include spills and leaks, even when there are no immediate health effects as yet (unlike contaminated land).

Of the many other statutes and regulations relating to the environment, attention is drawn to The Pollution Prevention and Control (England and Wales) Regulations 2000, which covers mobile plant and pollutants, and the Control of Asbestos at Work Regulations 2002 which relates to the need to identify asbestos, prepare and make available asbestos registers for each building owned, tenanted or managed.

The Party Wall Act 1996 enables work to be undertaken to walls shared by neighbours and for excavation to be safely undertaken in close proximity to adjoining owner's property.

Another environmental issue is that of energy performance (and in particular the Energy Performance of Buildings Directive) which is covered in greater detail in Chapter 9, but can be briefly summarised as introducing a requirement for new buildings to meet energy performance requirements. Energy performance certificates will be required before a building is constructed sold or leased. Boilers are subject to one of two regimes, either requiring an inspection with an energy report or a formal procedure for advising on efficiency.

More efficient buildings imply better quality and lower operating costs which should be reflected in negotiations and hence valuation.

Liability to specific third parties

Third parties originally enjoyed rights under common law, but statute law has developed to enhance these rights. Common law is not static and developments in statute law have often been followed by a wider interpretation of the common law.

The occupiers of premises fronting a highway are under a particular duty at common law to ensure that their property is not a danger to persons using the highway. Similarly, there is a liability to owners and occupiers of adjoining property; this may relate either to the condition or the user of the premises. A third and most important liability of occupiers is to persons coming on to the premises. If such persons are on the premises as a result of a contract to provide some form of services then the Unfair Contracts Term Act 1977 prevents the exclusion of negligence for business liability.

The Occupiers Liability Act 1957 applies to all visitors who come on to the premises with the express or implied permission of the occupier or in exercise of some statutory right. It imposes a duty on the occupier to take care to see that the visitor will be reasonably safe in using the premises for the purposes for which he was invited or permitted to be there. The position of trespassers was much improved by s 1 of the Occupiers Liability Act 1984. The occupier who has reasonable grounds to believe a third party may be in the vicinity of some danger, must take such care as is reasonable to ensure that he does not suffer injury. S 2 of the Act reverses an effect of the Unfair Contracts Term Act 1977, by enabling those who permit access to land for recreational or educational purposes to seek to exclude liability for any injuries.

The 1957 Act also dealt with the rights of visitors who had accidents due to the failure of the landlord to carry out his repairing obligations. While this was effective in respect of the common parts, it was not very effective in respect of the demised premises, due to the implied covenant that the tenant must have given the landlord notice of the defect. As a result this was replaced by s 4 of the Defective Premises

Act 1972. This provides that a landlord under an obligation to repair owes a duty of care to all persons likely to be affected. The duty arises not only when the landlord is aware of the disrepair but also when he ought reasonably to have known, and reserving a right to inspect would seem to imply the latter conditions as satisfied.

As between adjoining owners, problems can arise in respect of party walls, as regards support, weather protection and consequential defects and decay. In *Bradburn* v *Lindsay*, 1983, one of a pair of Victorian properties was demolished. The occupier of the remaining property claimed under negligence and nuisance he had lost support but more particularly dry-rot had spread from the demolished building and exposure to the weather was causing further decay. It was held that the plaintiffs were entitled to:

> have the exposed wall treated with dry rot prevention, recover costs of treating dry-rot, have the exposed wall buttressed and furnished in appropriate rendering.

The local authority can take the initiatives under s 29 of the Public Health Act 1961, as amended by the Local Government Act 1982. They can require a person undertaking demolition to:

> shore up any adjacent building, weatherproof surfaces of an adjacent building exposed by the demolition, repair and make good any damages to an adjacent building.

There is a right of appeal on the basis that the adjacent owner should contribute.

The Law of Property Act 1925, and in London the London Building Acts (Amendment) Act 1939, create a pattern of rights and duties, with provision for notices and the determination of disputes, the aim being not unnecessarily to restrict development, redevelopment and refurbishment.

The Access to Neighbouring Land Act 1992 provides for the first time a remedy against the unreasonable neighbour. The court will only issue an access order for works for the preservation of the dominant land, but the court retains an ultimate discretion to refuse access where this would interfere with the servient land to an unreasonable extent. The order must refer to the works, the area of the servient land and the period of time involved. The court may require compensation "to be paid and where commercial property is involved" consideration, where regard is had to the likely financial advantage to the dominant land. Of course the effect will be to encourage the parties to reach an agreement, with the hearing and costs as a last resort. In the case of *Sweet* v *Sommer*, 2004 the court was willing to imply an easement of necessity in favour of an adjoining landowner who could only have obtained another means of access by demolishing a building on his land.

Finally, the surveyor should also be aware of the Contracts (Rights of Third Parties) Act 1999 which changed the basic privity of contract rule in certain circumstances with effect on all new contracts entered into on or after 11 May 2000. Under the Act, a third party is able to benefit from a contract in either of two different circumstances:

- where the contract expressly provides that he or she may or
- where the contract purports to confer a benefit on him or her, unless "on a proper construction of the contract it appears that the parties did not intend the term to be enforceable by the third party".

The Act only confers rights on a third party and not any liabilities and such rights can only be excluded by a specific statement to that effect in the contract. Its most immediate effect upon surveyors may be in the contract setting out terms and conditions of engagement although it can equally apply to any other contract (eg cleaning or maintenance).

Employers

Quite distinct from their liabilities as occupiers, occupiers also have responsibilities in their role as employers. This responsibility has built up from various statutory provisions, starting with the Employers Liability Act of 1880. The Health and Safety at Work Act 1974 brought to an end the previous piecemeal approach and presents a framework within which a comprehensive code can develop, enforced by a single safety authority, the Health and Safety Executive. Failure to meet some requirements is treated as a criminal rather than a civil offence and one of the most important requirements is a safety statement putting on record the employer's policy and organisation for safety and welfare. The real teeth of the Act are the powers to make detailed regulations supported by codes of practice to assist in their effective implementation.

The two most important specialist statutory provisions are the Factories Act of 1961 and the Offices, Shops and Railway Premises Act of 1963. Both Acts adopt the widest possible definition of the types of premises which are covered. The requirements include: minimum working space, minimum temperatures, minimum sanitary conveniences, standards of ventilation and lighting, access, safety of machinery and specialist arrangements in connection with noxious or dangerous materials. The Fire Precautions Act 1971 originally applied to premises not covered by the Offices, Shops and Railway Premises Act and the Factories Act, but the Health and Safety at Work Act 1974 has applied this Act more widely. It is primarily concerned with the need for fire certificates and the means by which they may be enforced and consequential works for fire protection. The Act has a particularly severe effect on some of the older private hotels where the costs of carrying out the works far exceed the financial resources of the operator and significantly diminished the value of the property.

A designation order under s 1 of the 1971 Act requires buildings in which more than 20 persons are employed to have a fire certificate. The certificate contains:

- the use covered by the certificate, the means of escape, maintenance of the means of escape
- the means of fighting fire
- the means of giving warning, particulars as to dangerous materials.

A certificate may impose matters such as:

- ensuring escape is free from obstruction
- maintenance of equipment
- instructions to employees
- limitations on persons in the building
- anything else the fire authority consider appropriate.

The 1976 modification of the regulations place the onus of compliance with the certificate in the case of multi-occupied buildings with the owner, and by implication his agent. In practice, the granting of a fire certificate will be conditional upon the testing of alarms and precautions, safety equipment and the keeping of a log book.

A person in default (the owner) may, if convicted be liable to an unlimited fine and/or two years imprisonment.

The Fire Precautions (Workplace Regulations) 1997 impose further obligations and duties on the employer at a property where persons are employed, including the need to carry out a risk assessment.

Major changes in the fire regulations are likely to come into force in 2006 which will change all the previous legislation by doing away with the system of fire certificates and putting a further responsibility on the occupiers (and probably also the owners) to take into account fire safety and to provide detailed risk assessments.

Under the Health and Safety at Work Act 1974 if an inspector finds there has been a contravention of the Act be may serve an improvement notice (s 21) a prohibition notice (s 22) or take proceedings in the magistrates court (s 33). If the occupier wishes to appeal against an improvement notice perhaps because the cost of works would be unreasonably high for the resulting benefit then he can appeal to an industrial tribunal within 21 days. It may be possible to negotiate a solution with the inspector or the tribunal may cancel, affirm or amend the notice.

In *Westminster City Council* v *Select Management*, 1984, the council were successful in imposing improvement notices on the property managers in a block of flats. It was held that while each flat qualified as domestic premises; the common parts, lifts and stairs, were non-domestic premises and made available as a place of work for the use of persons who came to repair and maintain the premises. Accordingly, the person with control of the non-domestic premises, had a duty to ensure, as far as was reasonably practicable, that they were safe and without risks to health.

A prohibition notice is more serious, this can direct that certain activities cannot take place unless the specified works are carried out. An appeal does not suspend the effect of the notice.

The essential grounds for the two appeals referred to above is that it is not "reasonably practical" to comply with the inspector's requirements. This implies a subjective practical judgment seeking to balance cost against the level and frequency of the risk that the inspector claims is a breach of the legislation.

Since the Health and Safety at Work Act 1974, health and safety considerations have been further covered in such statutes and regulations as the Management of Health and Safety at Work Regulations 1999, the Workplace (Health Safety and Welfare) Regulations 1992, Control of Substances Hazardous to Health (Amendment) Regulations 2003 and, most importantly for the construction industry, the Construction (Design and Management) Regulations 1994, as amended by the Construction (Design and Management) (Amendment) Regulations 2000.

Various activities require licences from the local licensing justices or a government agency or need to be registered with the local authority. For example, to obtain a licence to sell intoxicating drink you must apply to the police, the local authority, and the fire brigade and provide details of the proposed licensee and the premises. In addition to full- and off-licences, a residential licence or a restaurant licence may be obtained, and all licences need to be renewed annually. The licensing justices also grant licences for betting shops and this involves the applicant in preparing plans of the area, identifying other betting shops and seeking to show the necessary demand for additional premises. Applications to the licensing justices may well be challenged by those already in the trade in that area.

Local authorities have now taken over responsibility from licensing justices following the Licensing Act 2003 and now grant liquor licences as well as being responsible for the registration of many specialist uses including scrap metal, pet shops, stables and employment agencies. Certain special agencies, such as the Gaming Board, issue licences to gaming clubs, and during 1979 Ladbrokes lost gaming licences for their most prestigious gaming clubs in London and this resulted in a fall in the price of their shares. The valuation of premises subject to the annual renewal of a licence is particularly difficult, and can only be undertaken after detailed research into the general nature of the trade, the standing of the operator or proprietor and the reputation of the particular premises.

Service providers — Disability Discrimination Acts 1995 and 2005

All service providers have, under the terms of the Disability Discrimination Act (DDA) 1995, a duty to make adjustments to the building (or provide a suitable alternative provision of the service) so that they do not discriminate against a disabled person or persons. However, it should be noted that it is the service provider offering the service, and not the building that has to comply.

The extent of the duties is governed by what is reasonable. In establishing what might be reasonable, the DDA refers to extent, cost, financial resource and disruption of adjustments. It should be stressed that even though a new building may be constructed based upon best practice and meet all current and proposed standards for disabled access, an occupier will probably not be meeting their obligations under the DDA if they do not have an effective management policy in place.

The provisions of the Act are being further extended by the Disability Act 2005 which will include from 2006 such matters as a widening of the definition of disability, private clubs and measures regarding letting of property. In this latter case, the original legislation only required disabled to not be treated less favourably — so a letting of the premises fell outside of the Act. The proposed legislation will create new duties for those disposing of premises to make reasonable adjustments to policies, practices and procedures. Examples of this might include:

- allowing a tenant with mobility difficulties to leave their rubbish in another place if they cannot access the designated space
- changing or waiving a term of a letting to allow a tenant to keep an assistance dog on the premises
- changing or waiving a term of a letting which forbids alterations to the premises so that a disabled tenant could make necessary access alterations, subject to consent.

Since most of the provisions of the original Act only took effect from 1 October 2004, it is still relatively new at the time of writing this book, and the courts have not yet had a chance to interpret some of the finer points of the law, which could have a wide implication. For example, one such cause for dispute could arise at rent review as to whether the cost of compliance (which falls on the occupier, as service provider) should be taken into account in fixing the reviewed rent.

Corporate social responsibility

From the above it can be seen that the measures described are part of the growing corporate social responsibility culture; they can have wide ranging effects on the property market. For example, high profile occupiers will not be enthusiastic occupiers of poorly energy rated buildings. Compliance with all the measures particularly in respect of any building works involved are likely to increase and any changes in ownership may trigger the need for an appropriate certificate which will become part of the due diligence processes. Questions will arise in respect of service charges, repairs and maintenance clauses in leases as regards liability and assumptions.

Commercial leases

The interpretation of leases lies at the very heart of urban property management. The estate manager is involved in the interpretation and implementation of covenants in existing leases drafted many years ago and also in advising on the drafting of new leases. The length of leases and the unpredictable changes in the requirements of the business community and of the property market are liable to result in problems quite unforeseeable at the time a document was drafted. A relatively small number of important cases have over the last few years played a major role in the interpretation of covenants, especially rent-reviews in times of rapid inflation.

Most leases are renewals of existing tenancies and negotiations will commence on the basis of the terms of the current lease, within the framework set out by statute law. In the case of a letting to a new tenant, the surveyor should, if he considers it necessary, obtain references from an appropriate source and, particularly when the prospective tenant is a relatively newly formed private company, a guarantor for the rent.

When agreement has been reached in principle for a new lease, heads of agreements are prepared by the lessor's surveyor, and the lessor's solicitor prepares a draft lease giving effect to that agreement and including all the normal lease covenants. Careful study of the draft lease by the surveyor to ensure its ability to give practical effect to the parties' intentions should ensure that at least the early years of the term should be trouble-free. Careful attention to all the terms of the lease is necessary to ensure that the terms have not been changed in the drafting and that the basis on which the rent was agreed has not been altered. The lessor will be anxious to ensure that the lease goes as far as possible towards what has been called "inflation-proofing" his income from the property. The lessee's negotiating position will be influenced by the extent to which it is a buyers' or sellers' market and the lessor's desire to implement a standard form of lease.

Lastly, attention should be drawn to the shortest Act on the Statute Book, the Cost of Leases Act 1958 which in its entirety consists of s 1:

> Notwithstanding any custom to the contrary, a party to a lease shall unless the parties there to agree otherwise in writing be under no obligation to pay the whole or any part of any other party's solicitors costs in the lease.

Thus, in the absence of any agreement to the contrary, each party bears its own costs.

This was considered in *Cairnplace Ltd* v *CBL (Property Investment) Co Ltd*, 1983, where the tenant successfully appealed against the judgment at first instance. The court had originally interpreted s 35 of the 1954 Act which covers the other terms of the lease as enabling it to make the tenant responsible for the landlord's cost in accordance with custom. However, the Court of Appeal held that statute overruled custom, since if it did not there would have been no purpose in enacting the legislation.

A code of practice for commercial leases in England and Wales — 2nd edn

There have been a number of reports and recommendations over the years such as the two Law Commission reports in 1985 regarding covenants restricting dispositions, alterations and changes of use. Since the mid 1990s there had been debate as to whether tenants were being offered sufficient choice and flexibility in lease terms and there was increasing disapproval by government of such terms as "upward only" rent reviews whereby the rent is assessed at review as either the current market rental value or the rent previously payable, whichever is the higher being imposed by landlords on tenants for a new lease. In the face of government calls for legislation, the RICS, the Law Society, British Retail Consortium, BPF, NACORE and DTI produced a voluntary code known as *A code of practice for commercial leases in England and Wales — 2nd edn* 2002 (see Appendix D). The code set out 23 recommendations relating to negotiation of a business tenancy and conduct during the lease. The most controversial of these was the suggestion that, although a landlord may seek upwards only rent reviews and full repairing an insuring terms, he should be willing to offer other options to potential tenants for other, less onerous terms, albeit for a revised level of rental.

It was monitored by Reading University who produced an interim and then a final (2005) report to the government as to its effectiveness. Whilst they found that upwards only rent reviews still dominated the UK lease market, because of the fall in lease lengths, the number of leases with no reviews had increased and was found in half of the new leases they reviewed.

As a result of this, the government has currently dropped its plan to legislate against upwards only rent reviews. The question of whether restrictions on rights to assign and sublet may still be the subject of future consideration by the government.

The demise

This should be clear and unambiguous with reference to appropriately coloured plans attached to the lease. When photocopying plans showing fairly complex physical boundaries, care must be taken that the copy is carefully coloured-up in the same way. It is usual for a schedule of rights and a schedule of reservations to be set out in detail at the end of the lease.

In certain very limited cases the tenant can increase the rights attaching to the occupation of the demised premises. The Law of Property Act 1925, s 84(1), enables the owner or the tenant, under a lease of more than 40 years of which more than 25 years have expired, to apply to the Lands Tribunal for the modification or discharge of a restrictive covenant. If the tribunal grants consent they may direct the applicant to pay compensation to any person entitled to the benefit of the covenant.

The tenant in using the premises may attach to them various chattels which then become fixtures. Following *New Zealand Government Property Corporation* v *HM & S Ltd*, 1981, it is necessary to distinguish between: fixtures put in by the tenant for the purposes of their trade, and those installed as part of the structure, which become landlords fixtures. The Court held that the rental benefit of the former could not, except by express agreement, become part of the demised premises and so were not improvements in the context of s 34 of the 1954 Landlord and Tenant Act.

Rent and guarantors

The rent, whether fixed or-stepped, and frequency of payment, on old or modern quarter days will be stated, together with the rate of interest on overdue rent.

Any rent free period for fitting out, any premium and any works should be stated with conditions which must be met. The service charge will be identified and its payment specified as additional or further rent. Although the detail of the rent review will be set out in a separate schedule, the dates for review, together with upwards/ downwards provisions, must be referred to, and it is helpful to require a memorandum to be attached to the lease after every review.

References will have been taken out in the case of less substantial tenants or enquiries made to establish that the tenant is indeed the claimed subsidiary of a "blue chip" covenant.

It may be necessary, in order to be able to recommend a tenant to the landlord, to obtain personal or corporate guarantees. The complexity of issues which may arise in respect of the liability of guarantors is illustrated by *Selous Street Properties* v *Oronel Fabrics Ltd*, 1984, the facts are complicated:

- Selous let to Oronel (of whom Morgan was a guarantor) in 1973 for 21 years
- Oronel assigned to Highlight Sports in 1973, who gave a direct covenant to Selous
- Highlight Sports assigned to Sunbird Fashions in 1976, who gave a direct covenant to Selous
- Sunbird Fashions assigned to Gavatina Holdings in 1976
- Karniol stood surety for the performance by Sunbird and with Sunbird for the performance by Cavatina.

Selous made a claim for £110,000 of rent arrears arising from a rent review in 1980. Oronel and Morgan argued that due to the effect of a licence given by Selous to Highlight in respect of retrospective consent for minor improvements they were excluded from their obligations. The court did not accept this since case law went against Oronel (the tenant) and while Morgan (the guarantor) would normally have been discharged by the variation, the detailed wording of the surety prevented this.

Whatever legal rights may be, their implementation depends upon the identification of a party able to discharge them. The judge had to consider rights of subrogation: he accepted that though the original lessee and assignee are both liable to the lessor, the assignee is primarily liable. Thus, Cavatina were primarily liable and so obliged to indemnify Oronel (Karniol was also obliged to indemnify Oronel). Thus, once Oronel had paid the arrears, they would be entitled to be subrogated to Selous rights. It was also held that as Morgan could not be in a worse position than Oronel he had a right to be indemnified by Gavatina and Karniol.

A tenant seeking to avoid these long-term problems has several routes open to him including:

- release on assignment (indeed *Pinemain* v *Welbeck International*, 1984, suggests that for an assignee to enforce a surety guarantee, the benefit must be expressly assigned to him)
- short leases

- options to break
- options to renew.

The problems encountered by a tenant when his assignee defaults have been forcibly brought home by the recession. In some cases tenants have found themselves liable for the debts of assignees many times removed down the alienation chain, whom they did not know existed and over whom they had no control.

As has already been mentioned, the Landlord and Tenant (Covenants) Act 1995 was introduced to enable tenants to limit their future liabilities under leases which they have assigned but it is not retrospective and so only relates to assignments of leases granted after 1st January 1995. It should be noted that the Act applies equally to landlords as well as tenants. Thus, a landlord will remain liable to the tenant under the landlord's covenants in a lease unless or until it is able to secure a release using the prescribed procedure under the Landlord and Tenant (Covenants) Act 1995. The case of *Avonbridge Property Co Ltd* v *Amit Mashru*, 2004 demonstrates the potential pitfalls, particularly with regard to personal covenants.

Turnover rents (whereby a tenant pays a set percentage of market rent plus a percentage of turnover from sales in a retail unit) present further potential problems, some of which will be discussed later. A particular problem that had to be addressed in the case of *Debenhams Retail plc* v *Sun Alliance & London Assurance Co Ltd*, 2004 related to how the turnover was actually to be determined. The lease, granted in 1965, defined "turnover" as being "... the gross amount of the total sales ... (by) the Tenant and ... any licensee". In 1973, VAT was introduced and the landlord queried whether the turnover certificate included VAT in the total turnover. The tenant confirmed this but soon realised that it was disadvantageous as they were now paying rent on a tax element (VAT) which was just passed on to the government. Although the tenant tried to challenge it with the landlord in 1976, 1990 and 1999, it was not until 2003 that it went to the courts to determine.

The court decided that it all depended upon the wording of the lease but in this particular case the judge felt that the turnover certificate would normally be produced on the basis of the tenant's audited accounts. These would provide a true and fair view of the company's profit or loss and would be produced in accordance with standard accountancy practices and would exclude VAT. Thus, the tenant was successful in claiming that the rent should be based upon turnover net of VAT. This had the effect in 2003 of reducing the rent on turnover by 17.5%. However, in 2005 it was successfully appealed by the landlords in their favour.

Options

The term of the lease is a matter of current practice.

In the past leases for 20 or 25 years have been common. As was reported by Reading University when considering the code of commercial leasing in 2005, due to a general reluctance of tenants to commit themselves for long periods, leases have generally become shorter, commonly 10 or 15 years, often enabling the tenant to break the lease at regular intervals. Longer or shorter leases can be granted, but the same objective of longer or shorter periods of occupation can also be achieved by options within a standard lease.

A lessee may seek an option to break the lease at intermediate date and this can include the payment of a fine; in the past, a lessor has usually only agreed to this when the building has proved difficult to let, since the effect on the investment value of his interest can be severe. However, American and European based multinationals are used to shorter lease terms, and invariably require options to break. The attractiveness of their covenant usually overcomes any hesitation the lessor may have on account of the uncertain long-term income, though this poses some interesting valuation problems.

Quite distinct from the statutory rights to a new tenancy, leases can contain an option to renew on similar terms. Landlords should beware of the risks of creating a perpetually renewable lease, which the Law of Property Act 1922 converts to a term of 2000 years. Where an option is given, the clause should specifically exclude the option provision from the terms of the new lease. Tenants seeking to exercise the option will need to ensure, in accordance with *Kitney* v *Greater London Properties*, 1984, that they have performed all the covenants.

This question as to whether the tenant has fully complied with their covenants has been considered on a number of occasions by the courts such as *Reed Personnel Services plc* v *American Express Ltd*, 1996 (outstanding dilapidations) and *Trane (UK) Ltd* v *Provident Mutual Life Assurance Association*, 1994 (breaches of lease — whether minor or not was irrelevant). In both cases, the option related to an option to terminate and not an option to renew.

A landlord's attempt to frustrate the tenant's exercise of an option was considered by the House of Lords in *Sudbrook Trading Estate Ltd* v

Eggleton, 1982. Here, the clause in an option (to purchase) provided that the price was to be agreed between valuers appointed separately by the landlord and tenant, and in default by an umpire, appointed by the valuers.

The lessors refused to appoint their valuer on the basis of precedent that as long as the agreement was incomplete the court could not order specific performance. The House of Lords overruled all the precedent since 1807 and ordered that the valuations be made and the property conveyed.

Upon the exercise of the option, the parties will be anxious to determine which terms, if any, can be altered and brought into line with current practice. In the early 1980s there were several cases on one aspect of options to renew, at the heart of the valuers expertise, concerned with the approach to a 21 year frequency of review. In *National Westminster Bank Ltd* v *BSC Footwear Ltd*, 1981, the Court of Appeal considered an option to renew for a further term of 21 years, "at the then prevailing market rent". At first instance the judge had accepted the landlord's contention that this could be interpreted as prevailing market conditions, and so the arbitrator could determine a rent for three or five years and provide machinery for the rest of the 21 year term. The Court of Appeal allowed the tenant's appeal and ordered that the arbitrator was to determine a single rent which was to be payable throughout the term of the new lease.

This decision was followed in *Bracknell Development Corporation* v *Greenless Lennards Ltd*, 1981, despite a bold attempt by the landlord to introduce the doctrine of frustration, as held to apply to leases in *National Carriers Ltd* v *Panalpina (Northern) Ltd*, 1981, on the basis of the impossibility for an arbitrator to determine a "full and fair" market rent for the whole period. The judge accepted expert evidence, not displaced in cross-examination, that an arbitrator would be able to determine a single rent for a period of 21 years in accordance with the clause.

These two cases appear conclusive law that no new rent review or variation can be inserted in the contractually renewed lease. In *Lear* v *Blizzard*, 1983, the court failed to award any uplift for a 21 year term, though this should not be taken as a general rule, but relevant only to facts of that case, a somewhat secondary investment; the decision has been subject to criticism.

The sparcity of the wording of the option clause often causes uncertainty as to how improvements are to be treated. In *Thomas Bates & Son Ltd* v *Wyndhams Lingerie Ltd*, 1981, where the court ordered rectification of a contractually renewed lease which failed to provide for an arbitrator to determine rent in default of agreement on review, the clause referred to "such rents as shall have been agreed between the lessor and the lessee". The court was able to distinguish this from *Ponsford* v *HMS Aerosols Ltd*, 1978, where there was no reference to agreement between the parties and so improvements could be disregarded.

An important lessor's option is a redevelopment clause. This provides that at specific dates during a normal occupation lease the lessor, on giving the required notice and if necessary being able to satisfy the courts, the lessor will be able to get possession for the purposes of redevelopment. This will not only depress rents but also limit the quality of the covenant. However, it does preserve some income from the property and avoid any empty-rate charges. What may be more important is that it preserves some reasonable economic activity in areas which can easily become blighted or derelict many years before the redevelopment project is commenced.

Repair

Of all lease covenants, those of repair are most capable of variety of phraseology. They may need to cover the standard in which the property is let, is to be kept, and is to be given up. Responsibility will need to be apportioned between the parties and procedures set out for the purposes of inspection and the giving of necessary notices.

The responsibility of the parties is not solely a matter of contract. The parties may find that due to the common law or statute they become responsible for a variety of unfortunate events that befall third parties as a result of the condition of the property, and adequate third-party insurance is essential.

In the case of a single occupier, a modern lease will usually be on a full-repairing basis. In the case of multi-occupation a variety of different patterns of obligations exist. Considerable difficulty has been experienced in law in distinguishing between repair and renewal. The case of *Ravenseft Properties Ltd* v *Davstone Ltd*, 1980, hinged on the interpretation of the tenant's obligation to "... repair, renew, uphold, etc. ...". The practical matter to be resolved was who should pay for replacing defective cladding at a cost of £55,000. The lessee's defence was that this was caused by an inherent defect. This was rejected by the court, who held that if an inherent defect relates to the whole building, then the lessor is responsible, but if related to a subsidiary part of the demised premises the lessee is responsible.

This should be compared with the case of *New England Properties plc* v *Portsmouth New Shops Ltd*, 1993, where the tenants were liable for a proportion of the landlord's costs of maintaining and renewing, the roofs foundations etc of the property. It was found that -the roof had been incorrectly designed and it had caused substantial fractures to a gable wall. The tenants were consulted and it was agreed that the original roof was inadequate. The new roof was different from the old in various ways. The court refused to uphold the tenant's submission that this work was in the nature of an improvement and the tenant had to pay his proportion of the repair cost.

Many phrases are used to indicate the obligation:

- in repair, in good repair
- in good and tenantable repair
- in good and substantial repair
- in structural repair.

Only the last of these contains a reduced burden. The standard of repair work required is that to reflect the state, condition and character of the property at the commencement of the lease. This raises the question as to the anticipated life of the repair on say a flat roof where cracks may be capable of remedy for a limited period by filling as opposed to a complete new covering. In *Manor House Drive Ltd* v *Shahbazian*, 1965, it was held that a patching repair was acceptable.

A tenant is not obliged to return to the landlord a wholly different property to that demised. Thus, in *Halliard Property Co Ltd* v *Nicholas Clarke Investments Ltd*, 1983, the court held that the covenant to repair did not extend to re-building in accordance with modern by laws, a jerry-built extension of 4 and a half inch brick work which fell down. However, it is worth noting that in the case of *Postel Properties* v *Boots*,1996, which related to repairs carried out to the roof of a shopping centre by the landlord and charged to the tenants by way of service charge, the court took a different view. One of the contentious matters related to the materials used. The court accepted the landlord's contention that the replacement of the original two layer felted roof by one made of three layers of felt was acceptable since

Another example is to be found in *Mason* v *Total Fina Elf*, 2003, a dilapidations case where tenant covenanted that it would "... to the satisfaction of the Lessor's Surveyor well and substantially uphold support maintain amend repair redecorate and keep in good condition the demised premises ...". It was held that the wording of the covenant presupposed that the item in question suffered from some defect physical damage deterioration or malfunctioning such that repair amendment or renewal was necessary. There was no authority for the proposition that purely anticipatory preventative work where no damage had yet occurred could be called for. Being old does not mean it needs replacing in anticipation of need.

In order to strengthen their hand landlords can introduce a tenants covenant, "to rebuild, replace, and renew whenever necessary in accordance with best modern practice, the whole or any part of the demised premises that may be or become beyond repair".

However, this may have a serious effect on rental value. In a rent review case, *Norwich Union Life Insurance Society* v *British Railways Board*, 1987, which concerned the rental value of an office building held on a 150 year lease the court upheld an arbitrator's award which made a downwards adjustment of 27.5% to reflect the onerous nature of such a repairing clause.

A prudent lessee should have a full structural survey carried out by an experienced building surveyor with necessary specialist assistance. This is equally true of a newly constructed building, and the surveyor should also be instructed to give an estimate of the likely repair cost of the building, and if this is in any way unusual, seek a commensurate balancing reduction in the rent. Instead of a full-repairing lease, a building may be let under what may loosely be called an internal repairing lease. This is usually the case in a modern lease of a multi-tenanted building where there are service charge arrangements. Also, on occasions older leases of single-occupier buildings were on this basis, and landlords wishing to alter the terms of the lease will have to satisfy the four tests in the *O'May* case, which were approved by the Court of Appeal.

Under internal repairing leases the main expenditure by the tenant is usually painting and decorating to a required standard at specified intervals, and in the last year of the lease the landlord will find it useful to be able to state the colour for the final year, since this may help a subsequent letting.

There are a number of points along the spectrum which allocate repairing obligations between the parties. In *Smedley* v *Chumley and Hawke Ltd*, 1981, the tenant covenanted to repair the interior and exterior and the landlord covenanted to keep the main walls and roof in good structural repair and condition. The foundations were found to be defective due to a lack of piling under part of the concrete raft. The landlord claimed the problem arose from a defect of design and that the remedial work would be an improvement of the premises originally demised to the tenant and so he was not liable for the repair. The court held that the only way for the landlord to meet his obligation was to carry out full remedial work.

In the case of a short-term letting, or the letting of premises in relatively poor order, particularly if they are blighted by some redevelopment or refurbishment proposals, then the landlord has to approach the matter in a very different way if he hopes to receive anything other than a nominal net income from the tenant, though even this would be better than nothing, since if the premises are unoccupied there may also be a liability to pay empty rates. A phrase such as, "to leave the premises in as good a state of repair as they were at the time of the letting, fair wear and tear accepted" will meet the interests of the two parties. There are many cases concerning the practical implications of this phrase and a useful summary is found in *Miller* v *Burt*, 1918:

> The tenant is responsible for repairs necessary to maintain the premises in the same state as when he took them. If, however, wind and weather had a greater effect upon the premises, having regard to their character, than if the premises had been sound the tenant is not bound so to repair as to meet the extra effect of the dilapidation so caused.

In *Regis Property* v *Dudley*, 1959, it was held that

> ... it does not mean that if there is a defect originally proceeding from reasonable wear and tear the tenant is released from his obligation to keep in good repair and condition everything which it may be possible to trace to that defect.

In the case of a fair wear and tear clause an agreed schedule of condition at the start of the lease, attached to the lease, will establish a bench-mark for the life of the lease. The surveyor should bear in mind that the document may not be referred to until the end of the lease and adopt an overall style appropriate to those circumstances. Problems can arise generally in respect of repairs when the lessor has covenanted to undertake work but fails to do so and it is a matter of urgency to protect the lessee's personal property. If the lessee, without giving notice, voluntarily does the work then in law the lessee cannot recover the costs, though there is nothing to stop his trying. If the lessee gives the landlord notice and the landlord fails to do the work, then *Lee-Parker* v *Izzet*, 1971, indicates that the cost of the work can be recovered by a reduced rent-payment.

If the tenant is responsible for internal repairs but there is no corresponding liability on the landlord for external maintenance, in most cases the tenant cannot compel the landlord to repair. This, and other anomalies was considered in The Law Commission Consultation Paper No 123 1992, *Landlord & Tenant Responsibility for State and Condition of Property*, but the matter has not been taken further.

Lessee's breach of covenant to repair

A claim for damages for breach of covenant to repair is usually brought at the end of the lease where, due to a lessee's breach, the lessor has suffered a loss in the value of his reversion. Assuming regular but not frequent inspections by the lessor's surveyor, the lessee should have been made aware of any failures to perform covenants during the lease by notice or interim schedules of dilapidations.

The provisions of the Leasehold Property (Repairs) Act 1938 provide relief against enforcement of the covenant to repair in the case of leases granted for a term of more than seven years when three or more years remain unexpired.

The Act provides that before commencing an action for damages, the lessor shall serve on the lessee a s 146 notice, to which the lessee may respond by claiming the benefit of the act.

The court is only to permit enforcement of the covenant if the lessor can prove one of the following:

- immediate remedy is necessary to prevent substantial diminution in the value of the reversion
- immediate remedy is necessary to give effect to legislation
- were the lessee does not occupy the whole property, immediate remedy is required in the interests of another occupier
- the breach is capable of immediate remedy at relatively small cost compared with the consequences of postponement.

A few months before the end of the lease, the lessor's surveyor, in accordance with the provisions for entry in the lease, should prepare a schedule of dilapidations in order to determine the extent of damages, if any, sustained by his client. The loss is the cost of carrying out the necessary work and any subsequent loss of rent. This provides an opportunity for the lessee to carry out the work or challenge the schedule. The lessor's surveyor should be clear as to his client's intentions as to the subsequent use of the building.

S 18 of the Landlord and Tenant Act 1927 sets a ceiling figure on any damages that may be recovered. The damages cannot exceed the amount by which the value of the landlord's reversion is diminished by the breach. Taken to its ultimate, if the lessor intends to demolish at the end of the lease or undertake substantial refurbishment, the breach of covenant will not have damaged his reversion, and no damages can be claimed.

The lessee's surveyor will need to advise upon the reasonableness of the schedule, both the physical items and the costs attached to them. The lessor's intentions in respect of the building, particularly the possibility of refurbishment, must be considered to establish the ceiling value to the claim. In the case of a fair wear and tear clause, the interpretation of the original schedule of condition is likely to create ample opportunity for a large difference of opinion as to the size of a claim for damages. Any attempt in the landlord's schedule to include inherent defects specifying work which will improve the building should be strongly challenged by the tenant. The case of *Fluor Daniel Properties* v *Shortlands Investments Ltd*, 2001, albeit relating to works carried out by the landlord and rechargeable against the tenant by service charge may be of interest in this respect.

The final result will usually be arrived at by agreement between the two surveyors and both will probably enter the negotiation with a little to give away. Should either adopt an inflexible attitude, there may be provision in the lease for arbitration. Alternatively, the matter will have to be referred to the courts, and this is probably a field in which the skill of the lawyers is least well suited.

Valuation surveyors will be required to advise upon the statutory ceiling value to the claim for damages. This requires an estimation of the value of the lessor's interest with a fully implemented repairing covenant, less the value of the interest in the physical condition as found, and it is only fair to say that such a calculation is as much a negotiation as is that between the building surveyors.

An increasingly common question is the extent to which the cost of the necessary work alone is recoverable. In *Drummond* v *S & U Stores Ltd*, 1981, where there was no clear evidence on the effect on the value of the reversion, the court awarded to the landlord a substantial part of the cost of repairs. The court also awarded three months loss of rent to cover the period for carrying out the works. The case of *Simmons* v *Dresden*, 2004, demonstrates a further difficulty which may face a landlord in establishing such a claim. In this case the landlord originally claimed a loss of £274,738.45 for making good the dilapidations plus interest. When it came to trial the landlord accepted that his interest had not been decreased in value and reduced his claim to £150,000. However, he had sold the property just over six months after the end of the lease for the high price of £2,655,000. The tenant argued that any dilapidation present had not reduced the sale price and so the landlord had suffered no loss. The court agreed and also observed upon the tenant's repairing covenant to keep the premises in "good and substantial repair"; the word "substantial" meant that its essentials, but not necessarily every last detail, must be kept in repair. This meant, in practical terms, that having regard to the age, character and locality of the premises, they were to be handed over in a standard fit for occupation by a reasonably minded tenant.

As an alternative to the preparation of a schedule of dilapidations and damages at the end of a lease, the lessor may exercise his rights under the lease to enter the premises and carry out repairs if the tenant has failed to execute them after due notice.

In really urgent cases the lessor may have no option other to simply take action under this provision. Lessor's surveyors must warn their clients that the costs are irrecoverable unless the full statutory procedures are complied with and this will not be possible.

In *SEDEC Securities* v *Tanner*, 1982, the lessors discovered stonework on the front of the building was falling on the pavement, they exercised their right of entry to carry out emergency work and subsequently served a s 146 notice requiring compensation, to which the lessess replied by claiming the benefit of the 1938 Act. The court held that a section 146 notice served by the lessor after the breach was remedied was invalid, and hence no damages could be claimed. However, in *Hamilton* v *Martell Securities*, 1984, it was held that where a tenant failed to comply with a repairing covenant in a lease which expressly conferred on the landlord the right to enter the demised premises, to carryout repairs and recover the cost from the tenant, an action brought by the landlord to recover such costs is a claim for a debt due under the lease and the 1938 Act does not apply. This was approved in *Colchester Estates* v *Carlton Industries plc*, 1984.

There is also specific performance. In the past the courts have been unwilling to order specific performance of repairs, due to the need for detailed supervision. In most cases damages alone were considered an adequate remedy, though in *Francis* v *Cowcliffe*, 1977, specific performance was awarded in respect of the replacement of a defective lift.

One of the most spectacular cases has been the problems of the New Scotland Yard Building in Victoria, the granite panel cladding of which was cracked. In *Land Securities* v *Receiver for the Metropolitan Police*, 1983, the lessor sued for forfeiture in order to obtain interpretation of the repairing covenant. The lessees sought declarations of the court that their proposals to replace granite with steel cladding would be an improvement; which had implications for the lessors. The lessees sued the architects, consulting engineers, developer and the GLC for negligence; the cost of replacement could affect the damages.

Improvements

Most leases contain an absolute or conditional covenant by the lessee not to make any alterations to the demised premises. Very often internal partitioning is permitted, as long as the premises are restored to their original condition at the end of the lease. This can be used to advantage by the lessor as the costs of purchasing and fitting partitioning are high and once removed the partitioning has little second-hand value. If the lessor waives his right to require the making good of the demised premises at the end of the lease, the lessee saves the cost of the work and the lessor can then let partitioned office space for more than open space. Only the more prestigious companies are prepared to have specialist-designed partitioning layout to meet their own occupational requirements.

The lessor may have good reasons for preventing alterations, for example, to protect the quality and features of the particular building or estate of which it forms part. Also, such activity by a tenant could have qualified as the commencement of a project of material development and create a liability on the lessor to pay any development land tax that a future government may impose with no prospect of additional income for many years after the date for payment.

The lessor may be quite willing to carry out alterations suggested by the lessee and in return for an appropriate increase in rent they can reach an agreement and amend the lease accordingly. It may seem easy to agree a certain annual percentage payment of the costs of the works. This should be approached cautiously, to ensure that the effect of rent reviews has been considered. Taking a long term view, the lessor's best interest would be served by taking a value approach and the lessee's by taking a cost approach. The arguments and factors involved were examined in *GREA Real Property Investments* v *Williams*, 1979, see section on the new tenancy, above.

Where a lessee wishes to make improvements, the 1927 Act provides for a procedure of notices and rights which balance the interests of the two parties. If the lessee does carry out the improvements then there is provision for compensation for the improvements on quitting. The Act also provides in s 19(2) that if the lease states that the works cannot be done without the lessor's consent, then that consent cannot be unreasonably withheld, though the landlord can obtain payment of a reasonable sum in respect of the diminution in value of the premises or other property in his ownership.

Lambert v *F W Woolworth & Co Ltd*, 1938, provides some detailed interpretation:

- the question of whether the works are improvements is to be considered from the point of view of the tenant
- generally, the onus of proving unreasonableness is on the tenant
- the court can determine whether the sum required by the landlord is too great
- alterations may be improvements if they enable the tenant to use the demised premises and the adjoining premises together more conveniently, although they may not improve the demised premises alone.

The question of when an alteration becomes an improvement is a matter that may cause a difference of opinion between the landlord and the tenant. For example, the conversion of an area of an office to form a kitchen and staff rest area which may be an improvement from the tenant's point of view; the landlord may merely regard it as a potential loss of office space.

Other problems that can occur are in the apparent delay by the landlord in responding to the tenant's request for consent to carry out improvements and in the landlord's refusal to grant consent. *Iqbal* v *Thakrar*, 2004, suggests that the principles for deciding whether consent has been unreasonably withheld for alterations same as assignment — but burden of proof is reversed ie it is up to the tenant to prove the need for the alterations.

Several statutes provide for the making of compulsory improvements, and can prevent the use of the property until these are carried out. Most of these powers are now covered by the Health and Safety at Work Act. Leases contain a clause placing the liability for this type of work on the lessee, but this may not be fair when the lease has a limited term to run and compensation on quitting may be totally inadequate. In, for example, the Fire Precautions Act 1971 the court is given power to apportion the cost of works equitably between the two parties. This does not mean it is necessary to go to the courts, since the parties may be able to reach agreement by negotiation. Apart from reducing the capital cost to the tenant, in due course it will also prevent him paying rent upon his own capital costs.

Insurance

Insurance has become an industry in its own right, the following are among the questions which need to be answered by the property manager.

Normal reinstatement basis, or some lesser amount based on indemnity where reconstruction of old and obsolete buildings would be uneconomic?

If normal reinstatement basis, have all the characteristics of the building and the factors influencing its reinstatement been considered, including:

- limited access
- site clearance
- architectural detail, changes in building regulations, statutes, bye laws, services.

What arrangements have been made to allow for inflation between valuations, and to reflect the period of delay, before completion of the reinstatement?

What risks are to be covered?

- fire
- special
- engineering and other perils
- loss of rent, reflecting reviews, for two, three or more years
- invalidation by tenant's breach of covenant
- terrorism.

What arrangements have been made to cover liability to third parties and employees?

Who is responsible for undertaking the insurance obligation?

How is the cost of premiums to be reimbursed and is there adequate exchange of information between interested parties?

Should VAT be included in the insured sum?

Is the insurance to be on a reinstatement basis or an indemnity basis?

Some idea of how badly matters may go is shown by *Beacon Carpets Ltd* v *Kirby*, 1984. The landlords covenanted to insure the premises, together with a sum to cover two years' loss of rent in full value in the joint names of lessor and lessee and that in the case of the destruction of the premises they would, with all convenient speed, layout all moneys received in respect of such insurance in rebuilding.

The landlords insured the building for £30,000 plus £3,000 to cover rent and fees, the policy named the insured as the landlords and the tenants for their respective rights and interests. The premises were destroyed by fire in 1977 and £26,484 was available for reconstruction, though reinstatement would have cost £50,000. In 1978 the tenants said they no longer wished to occupy any new building and the site remained vacant. In 1979 the tenants surrendered their lease to the landlord, the two parties shared the insurance proceeds equally and the landlord sold the site for £20,000. By their own acts they had released the clause from operation.

The Court of Appeal were somewhat unhappy at the way the parties had moved away from the pleadings at first instance. They allowed an appeal by the Beacon Carpet from the judge at first instance who had awarded £2 nominal damages and said that the insurance money belonged to the two parties in the ratio of value of their interests immediately before the fire. An increasing number of leases contain a clause of this type, that is to say, if the premises are not rebuilt within a specified time period (three to five years depending upon the building) then the insurance money is to be shared in the ratio of the value of the parties interests immediately before the event.

Service charges

The complexity and quality of environmental services in modern property, particularly air-conditioned offices and enclosed shopping malls can result in service charges being a relatively higher proportion of rental. Such charges then become an extremely sensitive issue between tenant and landlord, the landlord should remember that at the end of the day it is they who pay for the service charge — the tenant will look at the overall occupational costs and high service charge and statutory costs will result in less being available to the tenant to pay in rental. This will be particularly true where there is an over supply of any given property in the market. Historically, service charge costs have always risen faster than the rate of inflation. Since the 1990s "green" issues, EC directives and greater Health and Safety legislation have increased service costs also.

It is normal to have a covenant by the lessee to pay service charge as additional rent and a lessor's covenant to provide the services in accordance with a comprehensive schedule consisting of two parts: first, procedure for the apportionment of the charge; and, second, the full extent of the services to be provided.

Fixed payments have long been rendered impractical by inflation and tenants require greater accountability for what they are actually paying and therefore the apportionment of charges, all of which is easily aided by modern property management computer based apportionment. Large multi-let buildings with substantial services tend to have accounting procedures where the tenants pay a quarterly advance charge based on an estimate of costs (budget) or on the previous years actual expenditure. An annual reconciliation then takes place with the tenants being required to pay the balancing charge, or receiving credit.

The charge may be apportioned in different ways.

- Specific percentages in the documents; do they aggregate to 100%? Merit of certainty, with provision for change if circumstances vary, eg extension built.
- Floor areas with the use of an agreed measuring code — most popular and equitable. In shops it is possible to have some "weighting" formula so that large shops are charged less than their floor area proportion.
- Rateable values, in *Moorcroft Estates Ltd* v *Doxford*, 1980, there was a reference to the ratio of the rateable value of the demises premises to the rateable value of the whole. During the lease the assessments of various parts of the building changed. It was held that for the purposes of apportionment, the rateable value ratios should be those existing at the date each expenditure was incurred. The principle was also asserted in *Universities Superannuation Scheme Ltd* v *Marks and Spencer plc*, 1998 Not popular as several years can pass before correct rateable values are established, with complex adjustments necessary. Original tenant may assign. Use made of the services, logical but impractical.
- Volume — useful only where cubic capacity of demised areas varies considerably, most relevant for space heating charges.

- Site area — similar principle to floor area but translated to business parks with large scale landscaping, roads and open areas, security and management.
- Frontages — used in some shopping centres but not necessarily equitable for all services as depths may vary.
- A "fair and reasonable" proportion as determined by the landlord's surveyor; flexible allowing landlord free to adopt any methods of apportioning service charge; from tenant's point of view it may give landlord too wide discretion; what happens on change of ownership?
- Index linked service charges usually based upon an initial annual amount stated in the lease. In the case of *Cumshaw* v *Bowen*,1987 the landlord was entitled to change the index used after changes to the basis of RPI. Such a basis may not always be fair and equitable as a service charge that limits any increases in cost recoverable by the landlord to an index does not always imply that the tenant's liability automatically decreases when the actual costs rise or fall at a lower rate; see *Jollybird* v *Fairzone*, 1990.

In all the above methods care needs to be taken with voids, landlord's occupation and improvements which were highlighted in *Pole Properties* v *Feinberg*, 1981, where the court rewrote the lease.

In an article in *Property Week* entitled "Think Again on Service Charges" (7 January 2005), Edward Fenwick-Moore suggested that new methods should be considered for apportioning service charges, such as even by reference to footfall or volume of trade for retail premises.

Services, usually set out in a schedule to a lease, will vary according to the building or estate, its common needs, and the type of use. Each building has, therefore, to be assessed individually. A typical multi-let office building, for instance, could include the following services:

- insurance
- repair and renewals to common areas (external and internal)
- heating, ventilation, air-conditioning, all mechanical and electrical equipment, and fuel
- electricity to common areas
- cleaning and cleaning materials
- window cleaning
- floral and plant displays
- external landscaping

- lift repairs and renewals
- fire alarm systems and fire fighting equipment
- water charges
- rent, rates on staff accommodation (or local tax)
- building staff — manager, porter, receptionist
- managing agent's fees or landlord's costs
- consultants' fees — engineers, surveyors, accountants, auditors
- cost of compliance with statutory requirements
- creation of a reserve fund
- cost of enforcing covenants against other tenants
- security, including CCTV, car park and building access system.

A shopping centre may have all these services and more, including promotion and marketing, and the cost of running a traders' association. An industrial estate will probably have considerable fewer common services, typically:

- road repairs
- landscape maintenance
- external common areas — cleaning and litter collection
- drainage maintenance
- drainage pumps
- common lighting
- security
- caretaker
- common signage
- management fees
- consultants' fees.

Business park services will encompass both office and industrial items, weighted according to the control granted to the landlord.

In order that tenants' contributions for major items of expenditure, eg extensive redecoration, plant and machinery replacement, may be spread evenly and fairly over the life of the building, the creation of a sinking or reserve fund may appear to be an ideal solution, avoiding wide oscillations in yearly expenditure. The funds can be held with independent trustee to satisfy tenants that the accounting is being properly done. Yet, although the idea seems simple, the practicalities are not and usually cause more problems than they solve.

Some major tenants regard reserve funds as unnecessary, since they are prepared to meet a variable service charge on the basis of cost incurred, knowing that overall they will make the same contribution. Smaller businesses usually find such funds attractive.

Some landlords have concluded that the administration and tax implications are unacceptable or, at least, not worth pursuing. While the tenants will be anxious to ensure that contributions to the fund are treated as an expense in the same way as rent, the landlord will want to avoid taxation of the fund and its accumulation of interest. Unfortunately, neither position is tenable.

It seems that the tax system has no fixed provisions for a sinking fund, ie a fund for future expenditure, with its estimates and future provisions. The treatment of such sums varies from one tax inspector to another and there is arguably a need for some legislation or general clarification to encourage the operation of such funds. There are four main ways in which sinking funds are set up for tax purposes.

Where the contribution is expressed as additional rent in the lease, the landlord is likely to be taxed on receipt of that contribution — as normal income — but will only be able to set the expenses (to be incurred in the future) against income during the accounting period in which the expenditure actually arises. In other words, the landlord cannot claim a tax deduction for funds to meet future liabilities. The landlord may, therefore, carry a tax burden for several years before any expenditure is initiated on which relief can be obtained. This may not be attractive to a landlord and poses the question as to what happens if a sale takes place in the interim. It may be possible, depending on the circumstances at the time, to require a purchaser to set up a similar fund, at the existing level, with the vendor reducing the sale price of the investment by an equivalent amount.

If the fund is not referred to as additional rent, it is charged under a covenant and, strictly speaking, contributions should be taxed as a trade, under schedule D, which may be less attractive than schedule A, which will normally be the scale applied when the contribution is made as additional rent.

A trust fund could be established, and tenants required to make payments in accordance with the lease terms. The capital received would not be taxed, but the income resulting from the investment would be liable to taxation. The tenants may have difficulty in obtaining relief from tax for their contributions, since it can be argued that the money has not been spent but rather has been allocated for a purpose. On the other hand, the trust will offer the advantage of protecting the tenant from the landlord's dealings in the fund or insolvency. A fund in the form of a trust may well be subject to inheritance tax, although it is doubtful whether it was intended that sinking funds should fall within the parameters of this regime.

A separate management company could be established by the tenants, or possibly even the landlord, and could be treated favourably for tax purposes were it to have "mutual trading" status. Schemes of this kind have been established in residential blocks, but they appear to have been less favoured for commercial properties.

From a tenant's point of view, the desirability of making a sinking fund contribution poses a series of questions.

- Does the tenant have faith in the landlord or the managing agents?
- Where will the funds be held and who will benefit from any interest accruing?
- What will happen if it assigns the lease?
- What will happen to the proportionate balance of the fund at the end of the lease?
- How it can be assured that the level of contributions is correct?

The main problem is, of course, the unpredictability of future expenditure and changes in circumstances; there is probably little point in establishing a reserve fund if the landlord thinks a sale of the building is likely within the next few years or if several leases are going to come to an end relatively shortly.

Where a long-term view can be taken, the establishment of a fund could be appropriate, but it is probably still worthwhile ensuring, where possible, that funds are expended before the end of leases and that a separate reserve fund is set up thereafter with tenants who renew their leases.

Another alternative method that has been suggested by authorities such as Peter Forrester is the charging of an annual amount to cover wear and tear on the specific items (eg lifts). When replacement or refurbishment of the item is required, the landlord should have received sufficient from the tenants to pay for it. This has the advantage from the tenant's point of view of no large extra costs on the service charge and would appear to be fairer on tenants with different length leases since they all just pay for the years when they have had the benefit of the amenity.

Tenants may raise the following queries on the substance of service charges.

Does the schedule of services include improvements or enhancement, the cost of which should be funded directly by the landlord?

Is what the landlord wishes to include in the charge actually covered by the schedule of services?

In *Capital & Counties Freehold Equity Trust Ltd* v *BL plc*, 1987, it was confirmed that a tenant's obligation to pay service charges only arose from costs incurred during the term. The obligation was coterminous with the lease and, accordingly, the tenant could not be required to pay for services implemented and arising after the end of the lease. The landlord had sought to clarify whether the tenant's covenant extended to expenses which may have been expended or incurred during the term irrespective of whether they were or not.

If it is covered, is it necessary?

If it is necessary, is the procedure appropriate?

In *Bander Property Holdings Ltd* v *J S Darwen (Successors) Ltd*, 1968, it was held that the landlord need not adopt the cheapest method available in supplying the service (as noted earlier). Furthermore, as long as it is reasonable to do so, the landlord can elect to do a permanent job rather than patching up *Manor House Drive Ltd* v *Shahbazian*, 1956 (further confirmed in *Postel* v *Boots* case as mentioned previously). In *Mullaney* v *Maybourne Grange (Croydon) Management Co Ltd*, 1986, the court decided that the replacement in a tower block of wooden-framed windows (which required painting every four years) by double-glazed maintenance-fee windows, went beyond what was necessary for the purpose of effecting the repair. However, in *Broomleigh Housing Association* v *Hughes*, 1999, the court upheld the landlord charging all the tenants equally for replacement of all windows except for those top one flat in a residential block despite that tenants having replaced his own windows at his own expense previously and therefore not directly benefiting from the works for which he had been charged.

If it is appropriate, is the procedure carried out and supervised correctly?

If it is correctly carried out, is the cost incurred reasonable?

Finchbourne v *Rodrigues*, 1976 is the authority for the proposition that the landlord can only recover costs from the tenants if such costs are fair and reasonable.

Is the landlord obliged to obtain competitive estimates or accept the lowest quote?

There is no direct obligation on the landlord in respect of commercial property, although there is an implication that the costs must not be exorbitant. In the case of *Havenridge Ltd* v *Boston Dyers Ltd*, 1994, the court upheld the landlord's right to negotiate an insurance contract at arms length without the need to get alternative quotes.

Is the certificate of costs required sufficiently explicit?

The landlord's main concern will be the comprehensiveness of the service charges" clause. During the 1970s, lessors gained a commanding position on service charges in both the actual approach to the items to be included and the restrictions on the lessees' opportunity for effective criticism. Some leases contain a final phrase in the schedule of services provided, to include:

> provision of any other service or facility and the making of any other payment which may reasonably be required for the efficient running of the building, the comfort of the lessees and the efficient running of the service areas

together with a clause that the decisions of the lessor's surveyor should be final on any matter, provision, liability, or apportionment.

These sweeping-up or sweeper clauses do seem to be weighted in favour of the landlord but, on interpretation, the courts do seek to make a literal interpretation. *Mullaney* v *Maybourne Grange* derived from such a clause.

The principle is established in cases such as *Sun Alliance & London Assurance Co Ltd* v *British Railways Board*, 1989, that a sweeper clause is intended to cover future services not envisaged by parties when granting the lease.

It is not intended to make good errors in the original drafting; *Jacob Isbicki & Co Ltd* v *Goulding & Bird Ltd*, 1989, sought to determine whether sand-blasting of external walls was within the power of the landlord's repairing obligations even though that type of work was not specifically mentioned. The landlord sought to recover these costs under sweeping-up provisions which stipulated that:

> ... the landlord may at his reasonable discretion hold add to extend vary or make any alteration in the rendering of the said service or any of them from time to time if the landlord at his like discretion deems it desirable to do so for more efficient conduct and management of the building.

The court decided that the sand-blasting works were not of the kind originally anticipated in the landlord's repairing obligations and therefore the landlord could not rely on the sweeping-up provisions to recover the costs.

The most important case in this field is perhaps *Concorde Graphics Ltd* v *Andromeda Investments*, 1983. The lease provided for disputes to be settled by the landlord's surveyor, whose decision was to be final and binding on the parties. The tenant queried the inclusion of some items in the account, the cost of other items, including the managing agent's fee, and the basis for apportionment. The court held that while the landlord's managing agents had made a demand for the service charge on the tenant in that capacity, the function of the landlord's surveyor in the clause was essentially arbitral. Although he is the landlord's agent, he must act impartially and hold the balance equally between the landlord and the tenant, notwithstanding that the landlord is his principal and paymaster. His position is no more delicate than the architect required to issue certificates under the standard Royal Institute of British Architects' contract. The managing agents were, however, unable to perform this arbitral function, and, while the court declined to accept the tenant's whole proposal, it did point that the problem had arisen because the landlord had appointed a firm of surveyors to act as both its surveyors and its managing agents, and that, in the circumstances, the landlord should appoint other surveyors to fulfil the arbitral role.

In the case of a dispute over service charge and in the absence of any expressed provisions in the lease as to the procedure for resolution, there are three broad options available:

- the parties agree to refer the matter to a third party, to act as either an expert or an arbitrator
- where the tenant has failed to pay monies due, as the result of a dispute, the landlord can sue the tenant for the outstanding costs and the tenant can defend its action in court
- either or both parties can apply to the court for a declaration on the validity of the service charge demanded.

Disputes are professionally frustrating, administratively time-consuming, and financially wasteful. While the immediate goal will be to settle the dispute, the management surveyor should search diligently to find the cause, and add it to the list of points to be avoided when setting up the next agreement.

In the case of covered shopping centres, the tenants' association may provide a forum for comment and discussion of service charges, which is not always available in respect of other types of property. It is as well to remember that the tenants' perception of service charges is a cost of the service provided, the most obvious of these are the daily and weekly cycle of simple physical activities on site. It is not unknown for charges to be one-third or one-half of the rent and rising annually, presenting very real challenges to the most skilled of property managers. The most likely opportunity for limiting service charges is an agreed trade-off of quality for cost, though over a period of time this could damage the attractiveness of the investment. Over the long term this will result in pressure for low energy design and construction and in the short term for regular review of the energy budget, to avoid heating empty accommodation or heating to temperatures in excess of quite acceptable working conditions. Most landlords and tenants should be aiming to achieve value for money, rather than the cheapest option. The industry has produced its own guide to service charges entitled *Service Charges in Commercial Property — A Guide to Good Practice* which should be followed wherever possible.

User

There are two general kinds of user clauses, those that refer to the use of the premises in general planning terms and secondary clauses which relate to the way that use is conducted. Most leases contain a specific user clause, such as "offices" or "shop" and this will be further restricted in some cases, such as a shop to a particular trade, though this, as indeed any other clause in a lease, can be varied with the agreement of both parties.

Unlike improvement and assignment or subletting, there is no implied covenant that consent cannot unreasonably be withheld but in the case of a conditional covenant s 19(3) of the 1927 Act prevents the landlord from taking a fine or premium if he does grant consent although it does allow recovery of legal and surveyor's fees in connection with the application for consent to change of use..

Planning law may provide some guidance on the interpretation of the specified use but there is no presumption of a mirror image of interpretation between the two fields.

In *Anglia Building Society* v *Sheffield City Council*, 1982, the lease contained a clause restricting the user to "a travel and employment bureau and theatre ticket agency", consent not to be unreasonably withheld. The tenants, Alfred Marks (employment agency) wished to assign to Anglia Building Society. The landlords argued that a higher rent would be obtained from a "Class 1" retail user. The tenants argued that the premises were already in a service use. The court held, by analogy with *Bromley Park Garden Estates* v *Moss*, 1982, that the landlord was seeking a collateral advantage and that consent was unreasonably withheld.

The user clause must be considered with great care both at grant of a lease and at rent-review. Though the clause may suit the lessor's overall property management aims, restrictive user may depress market rental. To be fair to the lessee a rent review clause should state the specific restrictive user as a factor to be considered. If it is silent there is a presumption that it is to be included. Only if expressly excluded and there is provision for a more general user clause to be used for the purposes of the calculation should the lessor get the best of both worlds.

However, in the *Law Land Company Ltd* v *Consumers Association Ltd*, 1980, the Court of Appeal had to decide a conflict between a generally drawn rent review clause emphasising an open market basis, and a specific user clause to the particular tenant in occupation. The court held that it was not the parties' intention that the rent should vary according to the tenant's ability to pay as this would have been unworkable, and the open market basis prevailed with no reduction for the specific user. This may at first appear to be in conflict with the court's judgment in *Plinth Property Investments Ltd* v *Mott Hay & Anderson*, 1975 where a substantial reduction was given in rent at review because of a user clause restricting its use to offices used by ë consulting engineers".

User covenants tend to be of two types: those relating to hours of use and those to the general well-being of the building, the lessess or the lessor. Normal hours of business are usually specified and in office leases the delivery of goods is required to take place outside those hours. This is particularly true of furniture and partitioning, which can seriously inconvenience other users of the building.

The general well being of the other tenants can be covered by what are in effect private by laws relating to the use made of common parts, and specifically excluding their use, for example, as waiting areas or for display purposes. The building and the lessor's interest can be protected by covenants limiting the installation of specialist heavy business machines and activity likely to prejudice the provisions of the insurance policy. Lessees of office buildings who are involved internationally in certain specialist fields or in data-handling may need to work irregular hours, and it is better that this be expressly provided for than to rely on generous verbal assurances in the solicitor's office.

This leads into security or in a more positive sense, safety. Enclosed shopping centres have the most comprehensive and demanding requirements and thus are a good example in order to identify the principal factors:

- control of vehicular access — hours of use for the public and service vehicles
- control of shoppers at entry and in the centre. Control of retailers in the use of the premises
- control of vandalism and shop-lifting
- liaison with emergency services
- management of in house staff
- application of new technology; electronic pass cards, closed-circuit TV with recording facilities
- testing of emergency procedures
- supervision of plant and machine contractors dealing with fire-alarm, fire-prevention and fire-fighting equipment
- anticipation of special risks.

Assignment and subletting

It is possible to classify clauses as containing:

- no restriction against assignment
- an absolute covenant against assignment of the whole or part
- a qualified covenant against assignment of the whole or part
- an express proviso that consent cannot be unreasonably withheld (but note 1927 Act s 19(1)(a))
- a covenant by the tenant to offer a surrender to the landlord before assigning.

Each of these is worthy of attention in respect of the broader management implications from the point of view of landlords and the specific operational requirements of tenants.

An absolute covenant places the tenant in a very difficult position and the value of his interest is potentially subject to a very substantial discount. If the lease contains a covenant not to assign without landlords consent, then s 19(1)(a) of the Landlord and Tenant Act 1927 implies a term that this is not to be unreasonable withheld.

In *Bickel* v *Duke of Westminster*, 1977, Lord Denning said:

> I do not think the court can or should determine by strict rules the grounds on which a landlord may, or may not, reasonably refuse his consent. He is not limited by the contract to any particular grounds. Nor should the courts limit him.

In broad terms, the refusal must relate to either the character of the proposed assignee or to the effect of an assignment on rental or capital values. It has been held unreasonable to refuse consent on the basis that if a licence to assign was granted the lessor would lose a good tenant of another of his properties. Also, in *Killick* v *Second Covent Garden Property Co*, 1973, it was held to be unreasonable of the lessor to refuse consent because he feared there would be a change of use (contrary to a further covenant in the lease) since the consent would not preclude the lessor enforcing the user covenant.

Under the Landlord and Tenant Act 1988, where a lease contains the proviso that landlord's consent for alienation cannot be unreasonably withheld, the landlord has a statutory duty to consider a request for assignment within a reasonable time, and to show that any refusal was reasonable. Failure to do this could result in the landlords being liable to the tenant for damages. In practice, it can be difficult for the tenant to show that a landlord has unreasonably delayed giving consent although it can be a useful addition to the tenant's armoury in obtaining it. Nevertheless, in the case of *Design Progression Ltd* v *Thurloe Properties Ltd*, 2004 the timing of four months for giving consent to assignment was held to be too long. The court found that the landlord was pursuing a deliberate policy of obstructing assignment with a view to recovering possession; exemplary damages of £25,000 awarded to tenant.

Lessors have sought to avoid the reasonableness test by making a proviso that as a condition of the consent the lessee must offer to surrender his lease, this was accepted in *Bocardo SA* v *S&M Hotels Ltd*, 1979. However, in *Allnatt London Properties Ltd* v *Newton*, 1980, such an agreement was held to be void under s 38(1) of the 1954 Act.

The judgment was somewhat that convoluted, but it appears that a tenant wishing to assign should:

- offer to surrender in accordance with the clause
- if the offer is rejected, request consent
- if the offer is accepted, inform the landlord that the agreement is invalid
- then request consent.

One way a landlord can still try to restrict assignment without an absolute covenant is to introduce a restrictive user clause, this is likely to depress rental values.

Where a building lease for more than 40 years contains a qualified covenant against assignment or subletting then s 19(1)(b) of the 1927

Act provides that consent need not be obtained as long as the lease has seven or more years to run and the landlord is given notice in writing within six months of the transaction; this does not apply to leases granted by public bodies.

The problem of assigning a lease can become more onerous when the property is over rented and many tenants seek to get round this by subletting at less than the market rent. In the case of *Allied Dunbar Assurance plc* v *Homebase Ltd*, 2002 the tenant granted a lease at the passing rent but then gave the sub-tenant a side letter to compensate the sub-tenant for the difference between the passing rent and the market rental value. The landlord challenged this in court and it was held that the subletting was in breach of the headlease. It has been argued that such a clause in a lease which prevents a subletting at less than the passing rent should be reflected by a reduced rent in any subsequent rent review. There may be some consolation however in the subsequent case of *NCR* v *Riverland Portfolio No 1 Ltd*, 2004 which shows that in trying to sublet an over-rented property, the tenant may pay a reverse premium to the subtenant as a "sweetener" to get him to pay a rent equivalent to the passing rent.

Finally, an unlawful assignee, that is to say where consent has not been granted, may seek relief.

Forfeiture

The ultimate remedy available to a lessor when a lessee fails to meet the covenants in the lease is to exercise a right of re-entry. The courts, with the aid of statute, have sought to limit the use of this remedy by looking for any act by the lessor which constitutes a waiver, requiring particular procedures to be followed and granting relief.

If there has been a breach of a condition of the lease, then the lessor's right of re-entry arises automatically. If there is a breach of covenant then a right of re-entry must have been expressly reserved in the lease. The lessor can demonstrate his intention to forfeit the lease for the breach of a covenant by serving a writ for possession. The right of re-entry can be lost by some act which acknowledges the continuance of the tenancy. A demand for rent, or acceptance of rent due after the breach, with the knowledge that the breach has occurred, is the most patent form of waiver and stop-rent safeguards must be built into the accounting procedures of managing agents.

The doctrine of waiver needs to be interpreted in different ways for "once and for all" breaches such as assignment or alterations and "continuing" breaches, typically user or repairs.

In the latter case waiver through, for example, acceptance of rent applies only to past breaches. In *Cooper* v *Henderson*, 1982, the landlord continued to accept rent knowing that the tenant was using the property for residential purposes in breach of covenant to use them for business purposes only. The Court of Appeal granted forfeiture as this was a continuing breach. The landlord may lose his right to forfeiture through estoppel, where he makes some positive action to accept the breach.

For breaches other than the non payment of rent, the Law of Property Act 1925, s 146, prevents the enforcement of a right of re-entry unless the lessor serves a notice on the lessee specifying the breach, requiring a remedy if this is possible, and the payment of compensation. Only if the lessee fails to take action in a reasonable period of time can the lessor commence proceedings.

If the breach concerns a covenant to repair then the section 146 notice must give effect to the provisions of the Leasehold Property (Repairs) Act 1938. This requires the notice to state that within 28 days the lessee may serve a counter-notice claiming the benefit of the Act. In these circumstances, in leases of more than seven years with more than three years to run, forfeiture can only occur with approval of the Court, see the section on Repair, above.

This can be contrasted with the Law of Property Act 1925, s 147, which only provides relief in some limited circumstances in respect of internal decorative repair. The relief is claimed by a lessee on application to the court, which considers all the circumstances of the case, including the unexpired term. The lessor is under no obligation to advise the lessee of his rights under s 147. If a lessee fails to take the necessary action after the service of a notice in accordance with s 146 and the lessor seeks to enforce his rights of re-entry, the lessee can apply to the court for discretionary relief. This can be granted upon various conditions, including time-limits and the payment of the lessor's costs.

A subtenant can apply to the court for an order vesting in him the parts of the property that he occupies for the residue of the subtenancy, and so be protected from the misconduct of his mesne landlord which has resulted in the exercise of a right of re-entry.

In *Cadogan* v *Dimorvic*, 1984, the landlords had obtained judgment in a forfeiture action for breach of repairing covenants in the main lease. It was held that the court had jurisdiction under s 146(4) of the 1925 Act to make a vesting order for a new term of part of the accommodation occupied by a subtenant, even though the contractual term had ended before the application for relief.

In *GMS Syndicates Ltd* v *Gary Elliott Ltd*, 1980, the landlords alleged immoral use of the basement by the subtenants of the basement. Gary Elliott, the lessee of the ground floor and basement, who occupied the ground-floor shop, were granted partial relief from forfeiture by the court, restricting the order for possession in favour of the freeholder, to the basement.

In the case of non payment of rent, the lessor has a wider range of remedies. He can exercise distress on the lessee's goods or sue on the covenant or, as is more likely, seek to exercise a right of re-entry reserved in the lease. There is no need for a s 146 notice, but the common law requires a formal procedure to be adopted. This can be avoided by the lease stating that rent is to be paid without demand. The basis upon which the courts will grant relief for non payment of rents is found in s 38 of the Supreme Court Act 1981.

The right of the sublessee to protect his interest on threat of forfeiture of the head lease is found in s 146(4). The court may make an order vesting for the whole term, or part of it, the property comprised in the lease, or part of it.

It is common to find a provision in a lease for forfeiture if the lessee becomes bankrupt. However, the Law of Property Act 1925 s 146(10) provides relierand in effect a trustee in bankruptcy or liquidator has a year in which to find a purchaser and obtain the value of the profit rent. This was considered in detail in *Official Custodian of Charities* v *Parway Estates Developments Ltd*, 1984.

In conclusion, *Peninsular Maritime Ltd* v *Padseal*, 1981, has left the law on the enforceability of covenants after the service of landlord's writ for forfeiture in considerable confusion. It appears that the landlord cannot enforce the tenant's covenants but the tenant seeking relief from forfeiture can enforce the landlord's covenants.

Rent reviews

The impact of inflation together with the search for growth by the institutional investors has concentrated interest on the interpretation of rent review clauses. The 1976 and subsequent Blundell Memorial Lectures have made a major contribution in this field by providing a vehicle for the wide dissemination and discussion of rent-reviews and other lease covenants, identifying the close liaison necessary between lawyers and surveyors in their drafting and interpretation. The aim being to ensure that the clause is comprehensive in its content, conclusive in its interpretation and practical in its implementation

Use of the model review clause in its entirety is by no means universal, although many parts of it are incorporated in many leases. Major landlords and their advisers have their preferred forms of lease and review clause and they are often reluctant to change. Most of the sections of the model review clause work reasonably well, partly because there is now sufficient case law to decide what the words mean. However, the difficulties of drawing up a satisfactory clause can be illustrated by how the courts have interpreted clause 3(c). This requires the parties to assume that the demised premises "are fit and available for immediate occupation". Many similar clauses have the additional words "and use". Until the case of *Pontsam Investments Ltd* v *Kansallis-Osake-Pankii*, 1992, most surveyors understood the words to prevent the tenant's arguments that he should pay a lower rent to reflect the fact that in the open market he would obtain a rent free period to fit out.

The court decided that the phrase did not entitle one to assume that the premises had been fitted out by anyone. "Fit" meant "free from defects", not "fitted out". "Immediate occupation and use" did not necessarily require an assumption that the premises were ready for full beneficial occupation. It could mean occupation for fitting-out purposes. In short, the tenant could argue for a rent free period (reflected by a lower rent for review purposes) to taken account of the hypothetical fitting out period. As the law currently stands therefore the effect of the phrase is the exact opposite of its intention.

This case, and others concerned with interpreting rent review assumptions (see *City Offices plc* v *Bryanston Assurance Co Ltd*, 1993, on rent-free periods) show that the further is the assumption from reality, the less likely will the courts uphold it. There is a "presumption in favour of reality"; the tenant pays for what the tenant actually has.

In *Compton Group Ltd* v *Estates Gazette*, 1977, the court stated:

> The construction of the review provision is a question of law, which must be determined by the court in order that surveyors may know what is to be the subject-matter which they are required to value. It is not the function of the court to give the surveyors direction as to how they shall make their valuations that is to say what factors to take into account and what weight to give to them.

Whatever the phraseology of the particular clause in the lease, the four most important elements are frequency, basis, index, procedure and arbitration. These are considered below.

Frequency

Older leases, that is to say those dating from before the early 1960s, contained rent-reviews at far less frequent intervals than current practice, where five years is normal and under favourable conditions three years can be achieved. Assuming that five year reviews are the basis in the market, then all other periods, whether shorter or longer, can be regarded as non standard. Since the whole rationale behind the relatively low yield on property compared with that on Government Stock (the reverse yield gap) is the prospect of growth in income, then a longer rent review period offers the lessee protection from a higher rent at the normal review interval and in some cases has an additional value. Lessees have sought to reject this argument; retailers in particular say they must compete in the High Street and each year's trade must be judged on an equal footing. A higher rent based on the benefits of not having a review in five years' time has to be paid out of the current year's profits. To the extent that both cases have a ring of truth the estate manager is able to make either case on behalf of different clients with a degree of conviction. Taking five year reviews as standard there was evidence, in a rising market, of agreement or arbitration of uplifts on rent of 5% having been obtained on seven years reviews, 10–15% on 14 year reviews and 15–20% on 21-year reviews, in respect of good-quality investment properties in the first half of the 1980s.

In the latter case an uplift of 31% can be justified, applying equated yield techniques and projected annual rental growth of 5%. Various tables (most notably *Jack Rose's Constant Rent Tables*), graphs and computer programs exist to assist such calculations but what can be achieved on the day is often more a reflection of the negotiating strength of the two parties or the background of the arbitrator than of the strength of theoretical models.

Where the economy is in recession and rents are falling it is unlikely that many tenants will pay extra for a long review pattern. Indeed, as such patterns are often within a long lease a tenant will argue for a reduction in rent to reflect the onerous long term repairing liabilities.

The matter of adjustment for the frequency of reviews was considered in the case of *Basingstoke and Deane Borough Council* v *Host Group*, 1988 where it was suggested that the rent should be increased for each year of the review above (or below) the "normal" level by 1%; this being on the basis that there should be no reference to the change in value of money or the property and that the rent should merely be assessed as the additional rent that could be obtained if there were no provisions for review. Subsequent judgments based on this case have reduced the percentage differences even further for longer frequency reviews.

In *Pugh* v *Smith Industries*, 1982, an independent valuer was faced with a review clause which required a "full yearly open market rent", excluding the provision of the clause. He gave an award in the alternative, first assuming the rent would be fixed for the whole 20-year term at £36,750 and, second, assuming five year reviews (the normal market basis) at £30,600. The court held that the only possible legal interpretation was that of the first alternative; very much against the interests of the tenants. A similar conclusion was reached in *Safeway Food Stores Ltd* v *Bandeway Ltd*, 1983, where the court held that the assumed fixed-term rent was for 69 years.

However, in *Prudential Assurance Co Ltd* v *Salisburys Handbags Ltd*, 1992, there was an assumption that there were to be similar covenants in the hypothetical lease as in the actual lease except for the provisions for reviewing the rent. As this was a lease with 78 years unexpired the landlord argued that it should be valued as such but without review even though there were actually seven yearly reviews. It was held that the term to be assumed was that which the landlord might reasonably be expected to grant and the tenant to take of those particular premises in the open market at the best rent. This was a considerable victory for the tenant as the likely term was five years or so and no uplift in rent was likely. This case provides further evidence of the extent to which the court. will go to see that, despite any assumption, the tenant pays for what he gets.

Alternatively, the lessor can encourage the lessee to agree to standard review frequency and offer a premium or its annual equivalent in compensation. This can prove a useful course of action, enabling the new income to be capitalised at a lower yield, enhancing the investment value of the lessor's interest. This observed change in the frequency of reviews has resulted in a new type of clause, review of review, whereby in longer leases the frequency of the review will itself by reviewed at say 20 or 25 year intervals to the then current market pattern, with provision for arbitration.

By contrast, a poor investment property, showing signs of functional, social and financial obsolescence, subject to an FRI lease, may well command less rent on a long fixed-term review than on a typical open-market five year review basis.

Finally, there is the problem of a contractual, rather than statutory renewal clause, see s 6.4.3b. The wording is usually along the lines of: a renewal, for one more term only on the same terms as the existing lease excluding the right to renew, at open market rental value. Landlords have sought to insert normal frequency rent reviews, in order to more easily interpret open market value, when these did not exist in the original lease. The courts held in *National Westminster Bank* v *BSC Footwear*, 1981, that this was not to be implied and that the arbitrator should determine, "the then prevailing market rent", reflecting the long period until the rent could next be varied.

Basis

Most leases seek to establish a rent at review date at the then current open market value. Various phrases may be used; fair rent, reasonable rent, rack-rent, and they may be further qualified as to the status of a willing lessor and/or lessee. Any phrase other than "open market rental value" (or market rental value as the RICS Appraisal and Valuation Standards now defines it) can cause problems in interpretation.

Briefly, it can be said that the review clause, creates its own hypothetical lease and this is illustrated below.

1. In *Evans (Leeds) Ltd* v *English Electric Company Ltd*, 1978, the lease used s 34 in the definition of market rental but added the assumption that there was a willing lessee.

   The arbitrator found for a maximum of £515,000 pa and a minimum of £290,000 pa depending upon the interpretation of the clause. The court found in favour of the landlord, "the willing lessee will be unaffected by liquidity problems, government or other pressures ... In a word his profile may or may not fit that of English Electric Company."
2. Silence of the clause in the matter of improvements resulted in the House of Lords decision in *Ponsford* v *HMS (Aerosols) Ltd*, 1978, that the tenant's rent at review included the value of improvements carried out by the tenant which were not a condition of the lease. This was due to the rent being described as "assessed as a reasonable rent of the demised premises". In order to avoid this problem the review clause should specify either than improvements are to be disregarded throughout the lease or

identify a time period for which they are to be disregarded. If during the lease the lessee wishes to carry out extensive improvements which require a longer pay back period, then by agreement the license granted by the lessor to carry out the works could extend the lease or the period of time before which the lessor would be able to reflect the improvements in the rent.

A convenient shorthand has been developed whereby if the clause stresses agreement between the parties, a "subjective" test is to be adapted, having regard to all the factors which would influence the parties in a negotiation; whereas if the clause refers to a reasonable rent for the demised premises an "objective" test is adapted which results in a rent based on the premises as existing at the relevant date, including improvements.

3. If the lease includes any unusual or onerous covenants, specialist-user clauses or options to break it must be clear how these are to be treated as factors influencing the basis for the review. As discussed above in User, in *Law Land Co Ltd* v *Consumers Association Ltd*, 1980, the Court of Appeal faced with a conflict between a general open market review clause and a restricted user clause in the lease to the specific tenant, rejected any reduction in the open market basis as unworkable.

In *Plinth Property Investments Ltd* v *Mott Hay and Anderson*, 1978, the use was restricted to consulting engineers" offices and the landlord had an absolute right to withhold consent for a change of use. The court determined that at the review date the restricted value of the premises was £89,200 pa but that on normal lease terms, permitting relaxation to other uses with the typical covenant that consent was not to be unreasonably withheld, the rental value was £130,455.

A case of similar character, with the benefit going in the other direction, was *Bovis Group Pension Fund Ltd* v *GC Flooring & Furnishing Ltd*, 1983. The rent review clause referred to a letting for office purposes. In fact only two floors were used as showrooms and stockrooms and there was no planning permission for the building as a whole to be used purely as offices. It was agreed that the rent in terms of solely office use was £85,000 and in terms of the actual use of the property was £75,000. The court held that the arbitrator should disregard the actual use and assume the entire demise was used as offices. Also, in *Trusthouse Forte Albany Hotels Ltd* v *Daejan Investments*, 1980, the court accepted a clause which substituted shopping for the actual hotel uses.

In *UDS Tailoring* v *BL Holdings Ltd*, 1981, a restricted user to, "men's and women's bespoke and ready to wear tailors", was held by the court to merit a 10% reduction from unrestricted open-market value.

4. Rent-review clauses usually specify the term to be assumed remaining until the end of the notional lease; if the lease is silent, then this can only be the actual term remaining.

See *British Gas plc* v *Dollar Land Holding plc*, 1992. In general, unless the lease is very clear as to the assumed term to be valued, the courts have tried to keep the assumption as far as possible in line with reality. For example, in *Lynnthorpe Enterprises Ltd* v *Sidney Smith (Chelsea) Ltd*, 1989, it was held that an assumed lease "for a term of years equivalent to the said term" meant the original term of the lease from the actual commencement date, ie the residue of the lease remained to be valued. A similar decision has been made on the words "the term hereby granted" in *Millett (R & A) (Shops) Ltd* v *Legal & General Assurance Society Ltd*, 1985.

Pivot Properties Ltd v *Secretary of State for the Environment*, 1980, established that the parties would anticipate renewal in accordance with the Landlord and Tenant Act 1954, if on the facts such renewal would be possible. This resulted in the higher of the alternative rents of £2.925 million and £2.1 million being adopted.

All these factors were relevant in *99 Bishopsgate Ltd* v *Prudential Assurance Co*, 1984. The most crucial aspect was how to give effect to the rental basis of vacant possession in a building of 300,000 sq ft on a 98 year lease with seven year reviews.

Both parties initially valued on a floor by floor basis to arrive at a rent of £7–451m pa, the tenant then claimed three deductions; a 10% allowance for size, 20% for length of the lease and 16 months rent-free over seven years. The arbitrator had only granted the first of these with an award of £6–7 million. However, he had also made an alternative award which recognised the rent free argument, but spread the allowance over 14 years (the likely period of a subletting) in the sum of £6.065 million pa. The tenant appealed, and the alternative award was accepted by the High Court and confirmed by the Court of Appeal.

Index, turn-over, upwards/downwards

An alternative basis for review, particularly in the case of more unusual types of property or locations, is the use of an index. For example, greenfield campus-style offices or a hypermarket where there is no market evidence could be indexed to other property rentals such as the average current open market rent on several named buildings or an adjacent modern warehouse unit. Such factors will affect the yield on the investment. Rent-reviews can also be set at fixed sums for the future, and only hindsight will enable the parties to determine the merit of their decisions.

Rent reviews lead into indexation, but indexation is not just an alternative means of handling the effect of inflation within property investment, it can profoundly alter the whole pattern of operation of the landlord and tenant system. Profit rent and market rent become nebulous concepts, leading to very real problems in applying both landlord and tenant legislation and traditional property management techniques to the investment.

In France and Belgium rents are indexed to a cost-index with, in effect, a review to the true market at longer intervals. Each year the rent is adjusted to the cost-index, but every three years the tenant has a right to break. Thus if true open market rent is below the cost-indexed rent, the lessee's threat to exercise his option to break should encourage the landlord to revert to true open market rent.

In the UK, Slough Estates have obtained indexation to the retail price index. In 1983 the company offered its tenants the free option to convert to normal three year review market rents, less than half elected so to do. Specialist industrial premises, more in the nature of plant than true buildings, have been let on a retail price index, reviewed annually and reviewed to market rent every five years offering real value security, as long as the tenant's covenant is satisfactory.

The development of covered shopping centres with a controlled environment and sophisticated services and management, has presented an opportunity for turnover rents which are used widely in North America.

Typically, the rental consist of a minimum base rent, payable quarterly in advance in the normal way, with a turnover rent exceeding that figure. This requires a definition and authentification of turnover with varying percentages for different trades, for example:

• variety stores	2–4%
• food	2–6%
• electrical and household	6–7%
• shoes and jewellery	9–13%
• licensed property	9–15%

The advantages to landlords include:

- lettings can be finalised earlier
- annual growth
- awareness of tenants trading activities and a shared interest in the performance of the centre
- the effects of improvements and refurbishment can be realised sooner.

The disadvantage to landlords include:

- the need to develop new relationships between landlord and tenant
- potential problems with the Landlord and Tenant Acts
- potential problems with funding, although turnover rents are becoming more generally accepted.

It is also necessary to amend the normal lease clauses dealing with assignment, user and rent review. Even an assignment for a similar use can cause problems if it is assigned from a cheaper mass market retailer to an upmarket one. The latter will sell a much lower volume of goods but will have a much higher mark up; in such circumstances the percentage of turnover may well be increased for the assignee's business in order to maintain the same rental income for the landlord.

The growth during the later 1960s of upwards-only reviews certainly protected the investor's income but its fairness has been questioned. When a lease ends and a new lease is granted the rent will be at the open-market level irrespective of whether rents have risen or fallen. Is the rent-review simply a lessor's clause or part of an entire exercise to ensure that reality and fairness is present during the lease in the same way that the Landlord and Tenant Acts operate at the termination and grant of a new lease? In *Stylo Shoes* v *Manchester Royal Exchange*, 1967, the court in determining the terms of a new lease ordered that the review by upwards/downwards. This was followed in *Janes (Gowns) Ltd* v *Harlow Development Corporation*, 1979, where the judgment reads:

> I am not satisfied on the evidence before me that the insertion of rent-review clauses for review in one direction only have affected the rents which have been payable and accordingly I think that the appropriate course to take is to insert a rent review clause in this lease to provide for variation in either direction but to make no consequential adjustment to the rent.

However, in *Charles Follett Ltd* v *Cabtell Investments Co Ltd*, 1986, the court inserted an upward-only review as one had been present in the old lease.

The 1989–93 recession, with falling rents and very difficult trading conditions, brought the rights and wrongs of upwards only reviews into strong focus. Generally, however, tenants preferred to obtain other concessions from landlords, such as rent free periods or break clauses and few upward/downwards reviews were negotiated. But see section on commercial leases earlier and in 2005 there is evidence of over renting in the retail sector so the issue is again very relevant.

Procedure

Generally the rent review clause will require the landlord to trigger its operation by the service of a notice at a certain date or within a certain period in a particular form, and it may require a rent to be specified. Some leases require the lessee to reply within a specified period of time to the lessor's notice, otherwise the lessee is assumed to have accepted the rent specified in the lessor's notice.

Inevitably cases arose where a lessor's notice was served late. In 1977 the House of Lords made a fundamental change to the significance of time in rent-reviews. The effect of *United Scientific Holdings Ltd* v *Burnley Borough Council* and *Cheapside Land & Development Co Ltd* v *Messels Services Ltd* can be summarised as follows:

1. Time, *prima facie*, is not of the essence of rent-review provisions, so lessors may serve notice late, but the rent when determined has effect retrospectively.
2. The parties may make time of the essence in the contract and the courts will support this, otherwise time is not of the essence. An example of this is found in the "deeming" provisions in *Henry Smith's Charity* v *AWADA Trading and Promotions*, 1983.
3. In *Drebbond Ltd* v *Horsham District Council*, 1978, the lease provided that the lessor could require arbitration to determine a new rent by the service of a notice on the lessee within 3 months of the lessor's notice initiating the rent-review, "but not otherwise". It was held that this phrase had by implication made time of the essence and the lessor, being out of time, failed in an application that the matter be referred to an arbitrator.
4. By contrast, in *North Hertfordshire District Council* v *Hitchen Industrial Estate Ltd*, 1992, even where it was "a condition

precedent" for the review that a 12 month's notice be served by the landlord, this was not sufficient to make time of the essence.
5. The conduct of the parties can make time of the essence.
6. In a situation where there is an option to break at the same time as a rent review, time is to be of the essence of the contract, because the tenant might have exercised the option if he had been aware of the new rental. Despite ingenious arguments to the contrary the Court of Appeal confirmed this is *Al Saloom* v *Shirley James Travel Service Ltd*, 1981, and *Coventry City Council* v *J Hepworth & Son Ltd*, 1982.

It took several years for the full implications of the decision to be felt. This is illustrated by a series of cases seeking to test what period of delay might make time of the essence. This has been brought to a conclusion by *Amherst* v *James Walker Goldsmith & Silversmith Ltd*, 1983, where it was held that if time was not of the essence, the right of a landlord to trigger a review could not be destroyed by mere delay however lengthy. A very practical point, where there has been a long delay, is the equity of the landlord enforcing a clause for interest on back rent as well as the back rent itself, it depends upon the value placed on landlord and tenant relationships, see *James* v *Heim Gallery (London) Ltd*, 1980. If the lease is silent on the date when the back dated rent review becomes payable, then *South Tottenham Land Securities Ltd* v *R A Millett*, 1983, provides that it shall be the quarter day following the award.

If the time is of the essence, then it is necessary to consider whether the conduct of the parties meets the criteria set out in the lease.

In recent years the courts have generally tended towards taking a commercial rather than a semantic view of what is required from the parties. In *Nunes and Another* v *Davies Laing & Dick Ltd*, 1985 it was held that it was enough for a tenant to make it clear to the landlord that he objected to the proposed rent. The notice did not have to totally unequivocal.

A lease may contain a clause of both types, as in *Esso Petroleum* v *Anthony Gibbs Financial Services*, 1983, where a rent review notice could be served by the landlords 12 months either side of the rent-review date of 6 July 1978. If no such review took place the landlords could serve a month's notice, to review the rent at any time before the next set review in 1984. It was held that "subject to lease", "without prejudice" negotiations between the surveyors were not an exercise of the earlier fixed-date review.

Mistakes are likely to be costly. In *Centrovincial Estates plc* v *Merchant Investors Assurance*, 1983, a solicitor's letter offered the defendant tenant a rent from the review date of £65,000, which the tenant accepted. A few days later the solicitor telephoned to say the offer should have read £126,000, it was held that the contract had been concluded at the lower figure.

The greatest of care is necessary in the drafting of notices. The landlord's trigger notice must be unequivocally recognised as such, rather than the mere opening of negotiations. The tenant's counter notice must clearly state an election for arbitration rather than indicating some protest at the suggested rent or appearing to negotiate.

A perusal of case law will persuade the surveyor that whether time is of the essence or not will depend on the wording of the particular lease and a very slight difference can dramatically alter the result. The moral is to observe all the time limits in the review procedure.

A further problem that can arise is in the case of an over-rented property let on an upwards or downwards review pattern where the landlord has not initiated the review. Such a situation arose in the case of *Hemingway Realty* v *The Clothworkers' Company*, 2005. The lease said that only the landlord can initiate the review, but the landlord was reluctant to do so because it could well lead to a reduction in rent payable. The tenant claimed that it, too, had a right to do so, citing previous cases where it had been found that landlords could not frustrate the machinery of rent review provisions. In this case, the landlord was successful as the court held that the fact that only the landlord could initiate the review was not part of the machinery of the rent review provisions.

In other circumstances, tenants have sometimes successfully forced the rent review procedure by utilising arbitration provisions in leases which stated that in the event of agreement on the rent review not being reached within x months of the review date *either party* could ask for the appointment of an arbitrator to determine the review.

Arbitration

This had developed across the whole range of commercial activities as the civilised way of settling disputes, whereby the parties agreed to a third person in whom they have confidence to settle the dispute, and to abide by the decision.

Before dealing with the determination of a rent review by an arbitrator it is essential to distinguish the arbitrator from the independent expert. The arbitration is a legal proceeding under the Arbitration Acts and the arbitrator reaches a decision on the basis of evidence put before him sometimes at a formal hearing. He can call for the discovery of documents and he interprets the evidence. His decision is enforceable as if it were a judgment of the court. Though he is not liable for negligence, the court can set the judgment aside on the grounds of misconduct. The independent expert is appointed jointly by the two parties to carry out a normal valuation and to give his own opinion on the matter to be decided. He may have regard to evidence submitted, he may have a hearing and adopt what he considers the most appropriate procedure. His decision is not enforceable directly by the courts and he is liable for action for negligence, see *Palacath Ltd* v *Flanagan*, 1985. The independent expert may, if he wishes, proceed in the same way as an arbitrator, though the following procedures are specifically those of an arbitrator.

The rent review clause will provide for the appointment of an arbitrator, usually by the President of the RICS. Before accepting such an appointment the arbitrator should.satisfy himself of his independence, competence and ability to resolve the dispute, and this could involve correspondence with the parties.

In *Thomas Bates & Son Ltd* v *Wyndhams (Lingerie) Ltd*, 1981, the rent-review made no provision for arbitration in default of agreement between the parties. The court having satisfied itself that the omission was a mistake and on the particular facts, ordered rectification by the insertion of a provision for arbitration.

Once the arbitrator has accepted the appointment he must ensure that in any communication both parties are treated in an identical way. Having obtained confirmation from the parties that they accept his appointment, the arbitrator will issue directions to the parties setting out requirements and a timetable for pleadings, consisting of claim, defence and reply and arrangement for expert evidence and costs. The procedure of the hearing is entirely at the discretion of the arbitrator: if necessary he is able to obtain a legal opinion. When he inspects the property he should ensure that both parties are given an opportunity to accompany him.

At the hearing the parties may be represented by lawyers with surveyors giving expert evidence, or surveyors alone. The law regarding expert evidence is complex and was made clear by Mr Justice Megarry in *English Exporters (London) Ltd* v *Eldonwall Ltd*, 1973. The arbitrator can accept direct evidence, that is to say comparables that the surveyor or his firm has dealt with, but any other evidence should only be admitted if supported by authenticated details of the property and a copy of the lease, or alternatively agreed between the parties.

In *Land Securities plc* v *Westminster City Council*, 1992, it was held that another arbitration award was inadmissable as evidence. The logic of the court's argument in that case would lead to the conclusion that an independent expert's determination is similarly inadmissible. In a depressed market there are few lettings and sometimes those lettings are the subject of confidentiality clauses. The unavailability of third party awards combined with the lack of open-market evidence make the settlements of rent reviews more contentious. It is, however, generally possible to obtain details of "confidential" information by subpoena.

In *Segama NV* v *Penny La Roy Ltd*, 1983. The lease defined the market rent as meaning "yearly rental value of the demised premises having regard to the rental values current at the relevant time for similar property ... let with vacant possession." In accordance with *Melwood Units* v *Commissioner of Main Roads*, 1979, the court held that the arbitrator was entitled to hear evidence of rents agreed after the review date, though the greater the lapse of time the progressively more unreliable the evidence would become. The point was made that the same would be true of rent agreed before the relevant review date.

The court also held that where the clause refers to vacant possession it is quite proper for the arbitrator to have regard to evidence of rents agreed with sitting tenants, though the arbitrator may think it appropriate to adjust such rentals.

In reaching his decision the arbitrator must only have regard to the evidence submitted, though the weight he gives to evidence will be conditioned by his own experience, which was the reason for his appointment. In *Fox* v *PG Wellfair Ltd*, 1982, only one party gave evidence before the arbitrator, and in his mind the arbitrator rejected most of this. The court held that in doing this silently he was introducing his own evidence and as there was no cross-examination none of the evidence was tested. Lord Denning said the arbitrator had misconducted himself and the award was set aside.

In *Top Shop Estates Ltd* v *C Danino*, 1985, the arbitrator's award was set aside because he had made his own investigations without disclosing them to the parties and asking for their comments.

The RICS Guidance Notes suggest the award should contain the following:

- the parties and the subject property
- the arbitration appointment
- the agreement to submit to arbitration
- the date of appointment

- the method of appointment
- the instrument of appointment
- the issue
- the decision on this issue
- the decision on costs
- signature and date.

One of more of the following may be included:

- details of the preliminary hearing and procedures
- a reference to agreed facts
- the date and place of the hearing
- the attendance at the hearing or written representations
- a reference to the inspection of the property and any comparables.

A reasoned award should be given if one of the parties requests it; the arbitrator should accede unless he can think of no good reason for doing so, if both parties make the request, he should comply.

The publication of the award is closely related to the costs and fees. There are the costs of the reference, inclusive of all the costs incurred by the parties and the award, and the costs of the award itself which includes the arbitrator's fee. Prior to the issue of his award the arbitrator will specify how the costs should be paid and he will not release his award until his costs are paid. The rent-review clause may deal with the allocation of costs between the parties.

This is overridden by s 61 of the Arbitration Act 1996 which provides that the arbitrator shall decide on the apportionment of costs irrespective of any previous agreement on division made before his appointment.

The use of *Calderbank* letters has become common in arbitration disputes. One party may serve on the other an unconditional written offer. This offer which will include a figure for the rent and make provision for costs incurred up to the date of the offer will be headed "without prejudice save as to costs". The arbitrator will be asked to make his award as to rent and then hear submissions on costs when the level is known. At that point the *Calderbank* offer can be produced. From the tenants' view, for example, if his offer was higher than the arbitrator's award, he would have a good case for all those costs incurred by him and the arbitrator after the Calderbank letter to be borne by the landlord.

The Arbitration Acts from 1950 onwards were consolidated by the Arbitration Act 1996 which had provided an essential background upon matters of procedure and set out the relationship between the court and the arbitrator acting in a quasijudicial capacity. It was s 27 of the 1950 Act which gave the court power to extend any strict time limits. Its application to rent reviews was first considered in *Chartered Trust* v *Maylands Green Estate Co Ltd*, 1984, where the tenant failed to demonstrate the degree of undue hardship arising from a time of the essence clause necessary for the court to grant relief.

S 1 allows for an appeal on a point of law arising out of the award. In the absence of the consent of all the parties the court cannot give leave to appeal unless it is satisfied that the resolution of the legal issue, "could substantially affect the rights of one or more of the parties". In *Duvan Estates Ltd* v *Rosette Sunshine Savories Ltd*, 1981, the court held that the fact that the arbitrator had looked at facts and figures arising after the review date was only of marginal effect and did not intervene.

S 2 gives the court jurisdiction to determine questions of law which arise during the course of the arbitration. The court will only consider this if made with the consent of the arbitrator or concurrence of the parties and the decision might produce a substantial saving in cost and substantially affect the rights of the parties. In *Chapman* v *Charlwood Alliance Properties Ltd*, 1981, the court was not satisfied that all the hurdles could be jumped.

In 1992, approximately 13,500 arbitrators and independent experts were appointed by the President of the RICS, this fell to 8,150 by 2004. A surveyor involved in estate management should expect to be involved in arbitration from time to time. Though arbitrations are private some are referred to the courts as a case stated, usually to interpret lease covenants. However, in *Belvedere House* v *King*, 1981, one of the parties to a rent review sued (unsuccessfully) a surveyor acting as an independent expert.

PACT

The PACT Scheme has been set up and is operated jointly by the RICS and the Law Society to speed up the agreement of lease renewals. Under the scheme, the parties may agree to ask the court to refer all or some aspects to be decided by an independent surveyor or solicitor. Take-up of the scheme was initially rather slow and only 14 surveyors were so appointed in 2004.

In *A Code of Practice for Commercial Leases in England and Wales*, 2nd edn 2002, (see Appendix D) Policy Unit RICS, it suggests:

> Recommendation 23: The parties should take professional advice on the Landlord and Tenant Act 1954 and the PACT Scheme at least six months before the end of the term of the lease and also immediately upon receiving any notice under the Act from the other party or their Agent ...

The future of commercial leases

The effects of the recession on tenants in the early 1990s underlined the problems associated with traditional 20–25 year institutional leases. These proved to be inflexible, with tenants unable to dispose of an extensive liability often paying a rent at a level in excess of market rent and tied into a long term commitment, even when the lease has been assigned.

Various suggestions have been put forward as to how the problem could be eased. However, according to Reading University's research into the code of commercial leases in 2005 leases have reduced in length such that many are only for a term of five years.

The number of 20 to 25 year institutional leases appears to be diminishing and the trend for shorter leases appears likely to continue due in part to a number of factors:

- the increasing adoption of the international accountancy regulations which will require liabilities under leases to be shown on balance sheets
- the imposition of Stamp Duty Land Tax (SDLT) on leasehold properties whereby the tax payable relates to the rent payable for length of the term granted
- the increasing requirement by tenants for greater flexibility for which they do not want to be tied into long leases.

Future governments are likely to reconsider the question of whether to legislate against upward-only rent reviews and are likely to take a look at the seemingly inequitable restrictions on assignment or subletting imposed by some landlords. However, it can be argued that if both parties freely enter into such contractual arrangements, it should be a matter for the market rather than governments to control. In fact, while preparing this chapter, the BPF has announced that several high profile landlords are to sign an declaration that their tenants will be allowed to sublet space at a market rent.

The occupation of property is one of the many factors involved in running a business. The ability to adapt it or move from it is vital for a healthy business economy. The days have passed when a landlord could go to sleep for 25 years, occasionally waking up to serve a rent-review notice or dilapidations schedule. Nowadays, much more active management is needed, there is generally a higher turnover of tenants, and more frequent upgrading of services and plant. This has had an effect on rent and yields, there is little doubt that it will mean more work for surveyors.

Sources and further reading

A code of practice for commercial leases in England and Wales — 2nd edn (available on the Internet at various locations such as *www.commercialleasecodeew.co.uk* or *www.bpf.org.uk/publications*).

Business Tenancies (6th ed), Estates Gazette, 2002.

Donaldsons' Investment Tables, compiled by P Marshall, 1977.

EGi — web-based subscription service at *www.egi.co.uk.*

Encyclopedia of Landlord and Tenant Law, Woodfall, Sweet and Maxwell.

Health & Safety Executive website *www.hse.gov.uk.*

Insurance of Leased Premises, Adams and Sinclair Taylor, RICS.

isurv — web-based subscription reference tool at *www.isurv.co.uk.*

Monitoring the 2002 code of practice for commercial leases Crosby, Hughes and Murdock, (*www.odpm.org.uk or www.rdg.ac.uk/crer*).

Paying for our Past, DOE, 1994.

Rent Reviews, Freedman and Fogel, RICS/ISVA Clauses, 272 EG 496.

Service Charges in Commercial Property — A Guide to Good Practice, Property Managers Association, available as a pdf file at: (*http://www.propertymanagersassociation.com/files/2002729212512_Service_Charges_in_Com_Prop.pdf*).

The Reverse Yield Gap, Bowie, 267 EG 138.

Think Again on Service Charges, Edward Fenwick-Moore, *Property Week*, 7 January 2005.

UK Lease Structures in the Melting Pot, Bernstein and Lewis, 1992.

Valluation: Principles into Practice, W H Rees and REH Hayward, Estates Gazette, 2001.

Valuation: Special Properties and Purposes, Phil Askham, Estates Gazette, 2003.

Part 3

Positive

7 Estates Policy

Positive management

The nature of an estate reflects the aggregate effects of the interaction between the institutional setting of the property market, consisting of economic, social, technical and legal factors, and the aims and objectives of the proprietary interest.

The extent to which those aims are defined and the vigour with which they are pursued and monitored will determine the distinctive form and character of the estate, making it more than the sum of the various property interests at a particular instant in time. The time-scales involved, which vary from the long term implications of new ground leases on the freeholder's ultimate reversion to the short term practicalities of day to day management, require a flexible approach to decision-making. The range of issues involved, from the technical performance of components of construction to the impact of new government policy on the overall performance of an investment portfolio require breadth of view and the ability to assimilate and react to new situations.

Positive management is more an attitude of mind than just the effective exercise of suitably approved procedures. The achievement of the right balance between the efficient operation of existing procedures and fresh analysis, innovation and its implementation is what distinguishes the exceptionally gifted from the able and technically competent professional.

With hindsight, it is all too easy in a typical annual report to record what has occurred, identify the reasons for it and then make the broadest of statements regarding the likely course of future events, expressing hopes for the future and appreciation of the work of the staff. Perhaps this is the role of a chairman's annual statement and it would be unreasonable to ask for a more definitive form of analysis. Within any corporate body in which property is either the main resource or provides the environment within which operational activities are pursued, much more rigorous internal procedures will be occurring in terms of both management and accounting.

The phrase "positive management" can be used too easily, without the adviser or the client appreciating its full implications. In the context of property generally and investment property in particular, the phrase has only been in general use since the mid-1970s. In fact, positive management has been occurring on the estates of some of the old institutions for several hundred years but only in recent years has it been stated explicitly, almost as an article of faith. The word positive suggests an intention to contrast with some other form of management, hardly negative, since this would be too critical of what by implication had occurred before. Rather, positive management as an alternative to passive management, which might best be described as prudent inactivity. The history of property since 1970 is adequate explanation for the emphasis now given to positive management; this can be defined in terms of the application of management concepts discussed in Chapter 2 to the different types of estates, but of equal importance are lively, trained and perceptive minds concerned with the active implementation of the policy.

The Government Estate (Civil Estate and Defence Estate)

The concept of a national estate is capable of interpretation in many ways. It can be seen as an informative but passive property record or as the management of land, buildings and infrastructure under varying degrees of control by the state, in the best interests of the public. It involves the links between clear issues of political policy and complex administration and technical problems, some of which arise from apparently conflicting requirements.

The term "Corporate Real Estate" could be used for the Government Estate as the function is to deal with the day to day matters that arise within the running of the real estate portfolio. Whether the action is an acquisition, disposal, rent review, lease renewal or maintenance, the rationale behind the approach adopted will be driven by the strategy developed for the business and the real estate portfolio. In this case the business is the delivery of services and individual departments and agencies are required to be responsible for their estate portfolio and deal with rationalisation of this portfolio as part of this responsibility. This change of emphasis took place on 1 April 1996 and had a radical effect on the estate management functions with advice being given at that time from a central government body, the Property Advisers to the Civil Estate, now the Office of Government Commerce.

At the same time the concept of incorporating private sector funding became particularly relevant to the future of capital expenditure. This had been instigated by the Conservative Government of the early to mid 1990s and carried on by the Labour Government from 1997. The concept is known as Private Public Partnerships (PPP) embracing all appropriate forms of collaboration for the delivery of public sector requirements. These related particularly to transport, health, prisons and government accommodation. Within the scope of PPP the most appropriate vehicle for private capital expenditure is the Private Finance Initiative (PFI) which could take the form of total or partial private funding for the provision of toll bridges, tunnels and sections of roadway or transport systems. In the National Health Service (NHS) the programme of new build hospitals is almost entirely carried out by PFI with total private sector funding for the provision of hospital facilities and the complete servicing of this facility under a contract for between 25 and 35 years. Similarly with prisons the total provision of serviced prison accommodation for the same length of time. The concept has been extended to the wider public sector with school refurbishment being carried out.

Another significant change has been the outsourcing or privatisation of the estate management function within government departments. This covers the removal of in house servicing or maintenance in terms of managing the provision of facilities, ie cleaning, catering, security, messenger and administration. The most radical move was the sale of the whole of the estate portfolio of the Department of Social Services to Land Securities Trillium (LST) again on a long term contract arrangement which was then expanded by the acquisition of Employment Service accommodation. This has now been translated into the Department of Works and Pensions and LST provide all the serviced accommodation for the department. A similar operation has taken place with the portfolio of Inland Revenue and HM Customs and Excise with the acquisition by Mapeley providing a similar service.

The final stage in a study of the national estate is the identification of a national policy. Are the actions of government simply the aggregate effect of a series of *ad hoc* measures dealing with individual aspects of land and property that surface from time to time in manifestoes and legislation; or, is there some case for seeking to establish policy and administrative procedures which encourage a broad approach to the prudent management of land and buildings, irrespective of the nature of the proprietary interest? To some extent this is a political question, but aside from the emphasis which a minister could give to statutory initiatives, the concepts involved have not enjoyed much discussion outside a narrow legislative setting. However, in 1985 a paper was presented to the National Economic Development Office which considered the £10 billion pa spent on public sector infrastructure and public buildings. There is concern regarding inadequate regular maintenance and lack of strategy between central and local government.

In the post-war period alternative Labour and Conservative governments have resulted in regular and predictable changes to the setting within which the management, development and taxation of land occurs. The policies reflected by the 1947 Act, the 1964 Ministry of Land and Natural Resources and the Land White Paper of 1974 were all subsequently changed by incoming Conservative governments. In local government, town planning, though monitored by central government, operates in a very local way, and indeed more emphasis is being given to local plans and less to structure plans and their overall supervision by the Secretary of State. Some form of development value tax cannot be ruled out following the 2005 election.

Government departments

The importance of policy in the area of land administration was recognised by central government in 1977 by the creation within the Department of the Environment of a Directorate of Land Economy, responsible for advising the Secretary of State on land policies generally, and surveyors were appointed as senior civil servants with that directorate. Though small in number, they can be distinguished from surveyors seconded to the Department from the Valuation Office Agency of the Inland Revenue, who advise on the practical aspects of land valuation, and those located in headquarters or regional offices responsible for the implementation of statutory provisions and particular programmes. The Directorate's area of responsibility included land economy aspects of planning, housing, transport, agriculture, development agencies and property finance. Their activities varied from advising on the representation made by the various pressure groups in this field to the bridging of the not unknown gulf between policy and detailed implementation.

The most extensive and rigorous analysis of the role and performance of the DOE was formulated in 1980 in the form of "Management Information System for Ministers" (MINIS).

This showed that the majority of urban estates policy issues were located in the Inner-Cities Directorate and the Planning and Land Use Directorate, the latter having absorbed the Land Economy Directorate.

The Inner-Cities Directorate was concerned with the development and implementation of policy initiatives for inner-cities, including:

- industrial improvement areas
- enterprise zones
- links between commerce, industry and local initiatives
- housing aspects
- partnership and programme area policy
- urban development corporations
- analysis of urban trends
- management of research programme.

The Planning and Land Use Policy Directorate was organised into five divisions, dealing with 11 main activities including:

- policy and procedures for public sector land transactions relaxation of central control amendment to Crichel Down rules abolishing selective checking by DV advice to other government departments
- policy and monitoring of derelict land reclamation
- development control policy
- monitoring and streamlining the planning machine
- policy on supply on development land liaison with House Builders Federation monitoring features on county basis management of research
- winding-up of Community Land Act
- commercial property market monitor working of 1954 Act shopping policy liaison with URPI
- Property Advisory Group structure and activity of property market planning gain public/private development partnerships property market information

- Land Registers monitoring operation and land release
- Estates policy
  - disposal of public sector assets mechanisms of development effect of capital allowances on industrial property professional advice to Inland Revenue professional advice on local authority rental income
  - accounting
  - liaison with private sector including, BPF, RICS, Chambers of Commerce
  - professional advice on British Rail Property Board and to Ports Directorate
  - liaison with regional offices.

In 1985 a specialist Land and Property Division was formed under Christopher Howes (who was subsequently appointed Second Crown Estate Commissioner) which was somewhat similar to the Land Economy Directorate of 1977. These functions are now to be found in the Office of the Deputy Prime Minister (for the moment!).

Relationships

The function of national estate management is concerned with the relationship between the aggregate value of the return or benefit enjoyed by the occupiers and users of land and buildings and their external effects, compared with the real cost of the resources used in the provision and maintenance of real property. The available resources are a combination of those allocated by the public sector and those available through the financial institutions.

National estate management first requires some subjective political judgment on the social consequences of general economic policy and the discretionary influence exercised by government on the parties concerned with the management of urban property. Over recent years both the channels of communication and the perception of those involved has been much improved.

The second part of a public sector policy for estates is the creation and general supervision of a statutory, administrative and institutional framework conducive to the efficient management of estates, whether in the public or private sector. Statutory aspects such as planning and taxation are the most obvious and the most direct in their effect. Unless there is real evidence that long delays in the granting of planning consents result in better decisions than would otherwise be given, then there is a presumption in favour of quicker decisions.

Apparent abuses of the planning system also need to be considered as is shown by the Town and Country Planning (Compensation) Act 1985 which reversed the effect of *London Borough of Camden* v *Peaktop Properties (Hampstead) Ltd*, 1983.

The result of the case implied that under ss 165 and 169 and the eighth schedule of the 1971 Act, the owners of blocks of flats could make application for planning permission for an extension of less than 10% with the hope of a refusal. This would entitle them to compensation. Following the case the London Borough of Kensington and Chelsea received 20 such "penthouse" applications of a compensation nature.

In taxation there is concern that tax-planning considerations caused by the desire to maximise the net financial return have distorted the true resource implications, which is against the public interest. Both new and existing policies need detailed scrutiny to ensure that the intended objectives either do not cause negative side-effects or that the net effect is sufficiently positive to justify their introduction.

In the UK there has been a tradition of professional institutions operating not only rules of professional conduct but also policing voluntary agreements which might otherwise be regulated directly or indirectly by statute. This is nowhere better illustrated than by the fifth edition of the Appraisal and Valuation Standards ("Red Book"), prepared by the RICS. This sets out the basis for all published valuations. These also have implications for determining the extent to which a particular task can be seen to have been performed in a competent manner.

The committee's functions are as follows:

1. to advise and assist members on request concerned with the valuation of property assets of companies in meeting the requirements of the Stock Exchange, the City Panel on Take-Overs and Mergers, and company law and accountancy standards
2. to keep under continuous review standards of valuation and procedures necessary to keep abreast with public policy and attitudes apart from the requirements set out in (1) above
3. to initiate the preparation of guidance notes as and when appropriate on valuation standards and procedures related to property assets
4 to arrange for valuation of property assets which are published in the press (eg a prospectus for a sale of shares) or referred to in the Press, to be read; in the event of any apparent inconsistency with

any RICS guidance notes, to invite the member concerned to discuss the matter with the committee; to provide answers to public criticisms of valuation or valuers, as appropriate;

5. to be responsible for liaison with the Stock Exchange, the Institute of Chartered Accountants, and the City Panel on Take-Overs and Mergers.

So far, only in respect of the valuation of the assets of insurance companies has the Government considered it necessary to provide statutory regulations and the guidance notes go on to provide useful commentary on the statutory instrument.

Though rather less than 1% of estate management practice is concerned with the management of the concept of the national estate, the remaining 99% of estate management practice is the management of that disaggregated national estate. Within public sector estates there will probably be more recognition of the national perspective, but the pressures of good neighbourliness and socio-economic change linked to political sensitivity are likely to affect the larger private estates.

Research

The supporting role of the encouragement of innovation and research is important, but only if the impact of proprietary interests is given due weight both in the design of the research brief, and the interpretation of the results. There is a danger that research expertise and methodology developed within the disciplines of economics and geography and applied in the currently unfashionable field of planning may be used with too many implicit assumptions from the philosophy of planning and the mathematics of aggregation. The diverse and heterogeneous nature of land and buildings means there is no common vehicle or form for assembling information and helping to trace through from cause to effect. The usual research techniques require a considerable time-span for analysis and synthesis, during this time the market can change and with it the impact of particular parameters built into the original research brief, resulting in doubts as to the relevance of the project.

For example, two industrial buildings may appear identical; if one had been developed by a property company and the other by a local authority, their site selection, funding, taxation, management are all likely to be different, but in each case appropriate to the overall terms of reference of the proprietary body. The aggregation of such data masks rather than illuminates the distinctive features contributing to decision-making in both the provision and management of such property.

If research is defined as, "a systematic enquiry to test hypotheses and further the development of both theory and methodology applied both to the procedures involved in the profession (techniques, institutional processes etc.) and the substantial issues", then it is instructive to observe its interpretation by different sectors.

Only the largest 15 to 20 practices have their own permanent in house research capacity, though this does not prevent other practices from commissioning or undertaking research. Such research may be:

- to enhance their capacity to deliver services to clients
- to meet a particular client's need
- to enhance the development of their practice.

Research may well be seen as an element of public relations. There may be some unease between the time horizons of those requiring research and those undertaking research.

In the public sector, research in our areas of interest, has often been planning lead. During the 1980s there was a move towards crisper more prescribed, shorter time-scale tasks with a stronger estates emphasis particularly looking at:

- the inner-city
- derelict land
- industrial regeneration
- land supply pipeline
- the effect of the development control system.

Academic research was until the late 1970s distinguished by its inability to identify issues in urban property and translate them into good research projects for a variety of reasons. A combination of circumstances was required, and has occurred to change the climate and a research capacity is being added to the skills of the property manager, from within his own education through the graduate schools.

The role of the RICS is to create an environment supportive of research particularly in the fields of education and resources. This will enable surveyors to undertake research that will benefit the profession and for research to be easily translated into practice to benefit surveyors.

The Economic and Social Research Council's Planning and Environmental Committee was responsible for research into problems relating to the spatial structure, development and management of the socio-economic and man-made environment at national, regional and local levels in the UK, in other developed countries and elsewhere in the world. Land use studies and planning, housing and residential location are central to its remit. In its policy statement of 1983 it made specific reference to "the study of the land development process, land management and land values, the supply of and demand for housing, transport and recreational facilities and the influence of public and quasi-public agencies on the pattern of urban and rural development".

Subsequent changes in government have established different parameters and responsibilities with the main areas of influence coming from the Office of the Deputy Prime Minister and the Department for Productivity, Energy and Industry, DTI.

The publication of *The Land Market and the Development Process, a Review of Research and Policy* by the School for Advanced Urban Studies in 1978 was an important reference point.

In 1983 two important conferences took place, "Land Policy" sponsored by the then SSRC and "Land Management Research" sponsored by the Heads of Department of Surveying in Polytechnics. These, together with a widening of the available means of publishing work indicate a major change. The client need identified is now being matched by a response from the researchers.

The RICS has established three roles:

1. stimulating research in areas typified by major structural changes, such as information technology, client relations and changes in market structures
2. facilitating research by responding to technical issues and supporting the research community
3. communicating research, by, for example, promoting major conferences, topics at the 1993 Conference included:
   - the accuracy of property valuation
   - valuation in a non-normal market
   - the impact of changing commercial lease structure
   - the concept of risk in property investment
   - refinance in recessionary markets
   - local commercial property markets
   - trends in floor–space ratios
   - the impact of property on occupier performance.

In the last 10 years property research has become mature, reflective, predictive and robust. Though on occasions there is still a lack of realism in respect of the availability of data.

Individual estates

The formulation of estates policy requires good communication between those who hold or represent the proprietary interest and the professional estate manager. The policy need not be a plan in the formal sense of a series of phased consequential events, it could take the form of a position statement with several alternative but not mutually exclusive programmes. The role of policy would then be reviewing the emphasis to be given in the implementation of each programme. The client needs to be well-informed on the significance of the issues which must be decided and the alternatives available. He may also look for some guidance on the effectiveness of the implementation of estates policy.

Implementation refers to the process whereby a statement of policy subsequently has a direct and specification effect on the management of an individual property interest. This implies an operational structure providing responsibility and accountability, though to go further and suggest that a particular form is appropriate to a particular type of estate is neither sensible nor possible.

The training and academic disciplines from which the surveyor's skills have developed tend to stress the uniqueness of every interest in real property resulting in individual solutions to individual and the complexity of the business environment means that surveyors have to work more closely with other professionals and line management. This has encouraged surveyors to review the link between policy and the conduct of their work, and as a result in private practice the taking of instructions has become more formal and more rigorous reporting styles and monitoring procedures have been adopted.

The long term consequences of a particular property decision coupled with the indeterminate life of some estates, such as both the old and new institutions, means a view has to be taken, often on rather weak evidence, on the likely outcome of a series of related activities. In this situation the best way of maximising income in the short-run is usually inconsistent with the agreed long term policy. For example, the timing of the refurbishment of an office building and the most appropriate specification to realise the greatest net proceeds is a matter of very fine judgment; it can require intentional voids for some months or even years.

One estate's view of the short, medium and long term may differ substantially from that of another estate, indeed they may vary in respect of different properties within the same portfolio. Nevertheless, it is a useful guide to consider the short term to be a period of less than a year and particularly the period covered by the various procedures for statutory and contractual notices. The medium term is then a period of between one and five years, similar to the interval between rent reviews and the review is a useful opportunity for the parties to vary the terms of the lease by agreement. This would also be consistent with the timing of major reinspections and the implementation of changes to the portfolio in the light of analysis of the physical and financial performance of individual properties. In the long term, that is to say more than five years hence, it is possible for so many parameters to change that the best plan is the one which keeps open as many alternatives as possible and brings forward into the medium-term plan those with the greatest potential consistent with any overriding portfolio requirements.

The overall character of the portfolio is shaped by a complex interaction between functional necessity, institutional factors and the opinion of key management personalities. The functional requirements may be in terms of property as a resource of commerce or industry, or as a source of investment income. As the purpose of the proprietary interest change, so will the use made of the portfolio and the utilisation of individual properties. This will be assisted by an efficient records system designed to store, retrieve and manipulate data in the most convenient form.

Changes in the legal, social and economic framework within which property rights are defined and used can be made very quickly by a new government and have a major effect on the management of the estate. The patent form of statutory provisions means that they cannot easily be overlooked, though considerable uncertainty can exist as to their detailed interpretation. Social and economic change are less obvious and require some perception across a broad range of human activity in order to analyse their effect and consequences for the portfolio.

Due to the relatively small staff involved in the management of estates, individuals are able to have a direct impact on the practical management of individual properties. Flair and personal charisma have their role to play, which may or may not succeed in identifying or creating opportunities, but their successful implementation depends upon a positive approach to management, utilising a wide range of resources and techniques, drawn from both experience and innovation.

Sources and further reading

Depreciation of Commercial Property, Salway, CEM, 1986.

Destination UK: International Property Investment and the Role of Taxation, T J Dixon, M J Beard, K G Pottinger, M Brennan, A D Marston, College of Estate Management, 2000.

Mixed Use Urban Regeneration, Dixon and Marston, College of Estate Management, 2003.

Property Research, Schiller, 269 EG, Property Market Review, Supplement 32.

Property Research, Stapleton, Estates Gazette, 1989.

The Land Market and the Development Process, SAUS, 1978.

8 Professional Practice

Introduction

The exercise of estate management skills occurs within an environment shaped by the character of the organisation in which the surveyor works, the extent and quality of the information that is available and the nature of the relationship with the client. Statute and case law have in recent years resulted in a much more rigorous approach to the relationship with the client, in the taking of instructions and the consequences of both written and oral advice. Developments in technology have enabled information to be stored and handled in completely new ways, offering opportunities for more comprehensive and analytical forms of reporting to clients. The surveyor is therefore faced with not only new types of tasks but new ways of performing old tasks, and at the same time constant reminders of the need for the highest levels of technical competence and professional judgment. In meeting the needs of clients, whether in the public or private sector, surveyors are working in large organisations, with delegation and specialisation of function. No longer does an individual client receive advice from an individual surveyor; both now operate within some form of corporate structure. The giving of instructions, their acceptance, reporting and associated records are all more formal and less personal than in the past.

Clearly there are differences between the records required for operational as opposed to investment estates. The former could be seen as part of an integrated facilities management operation where cost per workspace or cost per unit of turnover or profit, are the key performance indicators. Whereas investment may constitute trading, development or income assets, each with different criteria for performance measurement, so property records need to capture and manipulate the necessary data to both manage the property and produce performance indicators which, together with relevant research, can be used to assess performance and plan the strategy.

The delivery of property management services in a professional office depends first upon the effective performance of three types of function:

- professional — knowledge, judgment, experience and personal skills
- technical — the performance of physical tasks
- accounting — financial records and action.

Second, upon the quality of communication between these functions. In a large organisation each task will tend to be carried out by specialised staff. Hence, the larger the organisation the greater will be the need for the arrangements for communication, interaction and monitoring. Major clients are likely to have framework agreements or call off contracts with joint access to the database. Thus, the role of the records system is not just to passively store information but to prompt approved action, monitor the results, and sound the alert when the outcome is unsatisfactory.

Information

Surveyors seldom get the opportunity to set up an ideal records system. Local government reorganisation in 1973 meant that some authorities needed a new system but immediate practical problems required immediate practical solutions. The new Unitary Authorities in the latter half of the 1990s are better served by asset management plans and asset rents. However, this created a patchwork, with a District joining a slice of a County, leaving an unbalanced county. The creation of an in house estates department in commerce and industry is another point in time when initiatives could be taken to determine an appropriate records system. Over the last few years these have increasingly been outsourced, after tending and possibly involving TUPE. In private practice the individual client's portfolios will be operated on systems with as many common features as possible. It is possible to make improvements to existing record systems and this will become necessary as commercial practices change, as the client's requirements become more sophisticated, or portfolios are split or merged.

An ideal records system can only be devised when a number of key questions have been answered.

- Who needs what information?
- Why do they need the information? How will the information be used?
- How quickly is information required and how is it kept up-to-date?
- How often is the information used?
- How will the various interested parties, lawyer, surveyor, accountant and client be involved in the development of the system?

As the answers to these questions become known, so preliminary conclusions can be reached on the appropriate means for the collection, storage and access of the information. It is important that the approach to information is dynamic, rather than static, that is to say the collection and storage of information should, while recording the situation at the date of initial entry to the system, emphasise the up-dating of data. Proformas for extracting information should prompt staff to seek out the correct information and capture new information as it becomes available whether contractual, commercial, financial, statutory or physical.

The kind of information needed for property management is extensive but the primary responsibility is financial accountability to the client. This requires monitoring the exercise of the responsibilities of the two parties and effective action on behalf of the client at certain key dates in the life of the property as specified in the covenants and by statute law. The property records, or what used to be called the "terrier" and is called the database in computer installations, contains the permanent details necessary to manage the property. A format suitable for the collection of this data within a landlord's management system will consist of sections covering:

- the property: address, description, accommodation, details of acquisition, services, user
- the landlord: name, address, account file number
- the tenant: name, address, account file number
- the lease: a summary of the terms of the lease and the implementation of rights under the lease.

The whole being supported by an extensive file of documents containing some, or all of:

- site plan and boundaries
- original architect's drawings and plans together with alterations
- specification and maintenance manual
- cost, date and extent of improvement expenditures
- planning consents
- photographs and plan showing position from which they were taken and date
- Ordnance Survey plans
- insurance policies
- leases
- other rights and duties in respect of the land
- other specialist material, consents and licences.

Depending upon the extent of the estate and the nature of the property, some form of summary will be available. It may be visual or geographically based, using Ordnance Survey maps or plans or overlays or simply a schedule, and it will in most cases be indexed primarily on the address of the property. The strength of a typical system based on Ordnance sheets is also its weakness. It usually takes only a few minutes to produce the record on any particular property but only good IT can answer:

- How many units do we own in excess of a certain size?
- What proportion of our space is vacant?
- How many rent reviews are in excess of seven years?
- How many units are occupied by a particular company?
- How many rentals are not quarterly in advance?
- Which leases have break clauses?
- With what company do we have the most insurance cover?
- How many leases expire in a certain year?

Occupiers

It is important to understand the role that property serves in the world of all corporate businesses — whether they are an international organisation with millions of square metres of office and production space across the world, or a small business. It is crucial for the real estate advisor to understand the nature of the client's business and the role that property serves — which is simply an *enabler* of the business.

The expression "the right property in the right place at the right time — and at the right cost" is often recounted by corporate occupiers — but in reality many factors conspire to prevent this perfect situation ever being achieved.

Historically the drivers of the location of corporate property varied, but could include proximity to:

- marketplaces
- clients/customers
- suitable numbers of staff with appropriate skill sets
- transport infrastructure
- like-minded businesses.

With the advent of the Internet and electronic communications, combined with (especially in the south of England) the lack of any numbers of suitably qualified staff in the "traditional" business locations — the Thames Valley is a good example — the pressures are different, and are bringing about some dramatic changes in the approach to corporate real estate. For example, today it would be very unusual for a query to an insurance company being taken by somebody in their "head office" in London or Edinburgh. It will almost certainly be taken by somebody sitting in a call centre in the far North East or South West of England — or India.

As a result of a shortage of suitable staff in many locations, as mentioned above, property in recent years has also been required to fill another role — that of helping to attract and retain staff by providing a modern image and better environment than perhaps the competition — and in the UK we have some of the most sophisticated headquarters office buildings in the world. This is another example of property as an enabler of the business — and in this context it is important to bear in mind that in a typical corporate, total property outgoings represent about 6%–8% of all running costs,[1] whereas staff typically represent 60%–70%. Nevertheless it is still a significant outgoing in absolute numbers, and is perpetually under scrutiny by finance directors as well as shareholders.

---

1 IPD Occupiers' Property Databank.

There are now several organisations who offer benchmarking advice in order to ensure that costs are of the right order, within any given sector. As referred to above, probably the leading organisation within the UK is Occupiers Property Databank who, in return for a significant membership fee, can offer organisations detailed analysis of occupational costs in various sectors.

The brutal reality is that the perfect scenario of everything being in the right place at the right time is very difficult to achieve for a typical corporate, mainly because of the *inflexible* nature of the UK commercial property market. The best example of this is the length of typical commercial leaseholds; while they have reduced from typically a 20 year term 30 years ago, to something approaching 15 years now, that should be compared with the fact that many businesses find difficulty in planning more than a year ahead!

This inflexibility does cause great difficulty in estate planning; as a result, in recent years we have witnessed some attempts by the property industry at large to address the problem.

In the early 1990s came the advent of the "serviced office" organisations who particularly targeted new/small companies who did not wish to enter into long leases as, due to their very nature, it was difficult to predict how much space they would need in years to come. The serviced office operators (major examples being HQ and Regus) offered enormous amounts of flexibility — being able to lease quality office space for a period as short as a week, up to a period of three years, complete with furnishings, shared reception, IT facilities etc — in return for an "all inclusive" figure, deep within which, if one cares to analyse it, is a rental figure equivalent probably to three or four times the appropriate market rate. Flexibility always equals additional cost. A conservative estimate suggests that this sector comprises approximately 900,000 m^2 of office space, representing around 0.8% of total office accommodation within the UK.

On a very different level, the mid 90s also saw the emergence of property outsourcing/real estate partnerships (REPs). These are simply a vehicle whereby a third party service provider (current examples being LandSec Trillium and Mapeley) take a "virtual assignment"/acquisition of all property and liabilities (with beneficial balance sheet implications) and in return for a suitable profit margin, enter into an agreement whereby the occupier may quit a given percentage of the estate over a pre-agreed period of time. The service provider will also, as required, provide all other property services ranging from structural repairs to reception staff. Such arrangements have been entered into by several larger government departments and private sector organisations such as the Department of Work & Pensions and the Abbey National Building Society.

One result of the recent outsourcing/REP phenomenon is that many corporates have discovered that the level and accuracy of information held on their properties within house is extremely poor. In order for any outsourcing organisation to be able to price their offer to a potential client, it is crucial for them to have, in great detail, all the relevant information on the property involved — in many cases, nearly always leasehold — whether it is a retail kiosk on Aberdeen station or a London headquarters building.

The outsourcing company/REP cannot make their bid without a full understanding of all the liabilities; many large corporates — both private and public — have discovered that they simply don't have the information available.

In response to this problem, several property oriented software houses have reacted to the opportunity and there are now a variety of electronic property databases available including Manhattan, TRAMPS etc. These systems can now be obtained off the shelf or designed on a bespoke basis, and, if properly used, can be an invaluable tool especially in the area of cost analysis and budget preparation. Most now are PISCES compliant for compatibility. However, on the premise of "rubbish in, rubbish out", relatively sophisticated IT skills are required to input the property data correctly especially where the data is required to interface with a financial forecasting model.

In the banking/financial services sector, characterised by the closure of hundreds of high street bank branches in the last few years; the banks have argued that this happened as a result of the ability of their customers to communicate on-line — although it also has the added benefit of reducing their exposure to property. This has resulted in significant disposal programmes by several of the well known banks and typically, a traditional high street bank branch in a regional town or city is now a wine bar — in response to a different demand altogether. In addition, nearly all the UK high street banks have outsourced their real estate function in response to the current trend of "sticking to the core business" — typically 40–60 in house surveying/legal/ architectural posts, and a contract for real estate services being awarded to a firm of advisors for a three to five year term, the theory being that the external advisor can respond more quickly, more flexibly, and with a better market knowledge based on a much broader skill base.

By way of contrast, the explosion in activity by the mobile phone industry over the last 10 years is very different. From the embryonic organisations of the mid 90s, perhaps typically being 2,000 m^2–3,000 sq ft of office space in their regional locations of Newbury (Vodafone) or Bristol (Orange), these companies have seen enormous growth due to customer demand, and now have headquarters buildings of 30,000 m^2 –40,000 m^2 together with several major call centres (typically the North of England where the workforce is available) and a retail branch network of 300–400 units around the country — and still growing. This very rapid and significant growth has, in itself, put pressures on the way the property industry responds to the needs of corporates — forcing it to react more quickly and flexibly, and significantly reducing timescales between agreeing transactions and getting people on site.

The warehousing/distribution sector has also undergone significant change in the last decade. Ten years ago the major retailers (key players in the distribution network) would have had a number of distribution depots located in medium sized towns to provide "just in time" deliveries to their surrounding stores. However, recent development has been for super sized regional distribution centres in excess of 50,000 m^2, located in a central England location, providing regular bulk deliveries to much smaller local urban distribution depots, in order to service a smaller number of retail outlets. The property may be just part of an outsourced service level agreement with a logistics supplier.

All local authorities were required from 2000 to produce asset management plans in order to "optimise the utilisation of assets in terms of service, benefits and finance return" or in operational terms, "knowing what properties you own, knowing why you own them and knowing you get the best out of them".

The key elements of this are:

- service need
- corporate property strategy
- corporate asset management group
- property reviews
- disposal plan
- maintenance plan
- data management
- capital programme.

The corporate asset management group needs representatives from estates, financial and the largest service departments, typically in a unitary authority, education and social services account, for 90% of the space.

The property reviews start from the needs of the services, reflecting changes in legislation, function and reorganisation, with the growing emphasis on joint provision.

Maintenance plans need to be driven more by best value rather than the unachievable pursuit of an aspired standard. The capital programme needs to be more than the total of isolated projects, rather to include a road map with options.

All require integrated data management of the widest range of property from public conveniences to shopping centres and from access rights to freeholds.

The core data comprises:

- property: ownership, tenure, tenancies, lease schedules, rights
- building: the capital asset, condition, services
- management: records and accounts
- contractors
- service departments.

This needs to be assessed by activity programmes to produce reports:

- diary dates
- works orders
- rentals received
- service charges
- valuation
- performance indicators of both properties and staff.

There is no doubt that the asset management plan process has increased recognition of the capital and revenue importance of property as a corporate asset, with service departments seeking value for money occupation.

The merger of the health/welfare functions of the community health trusts and social services departments of unitary authorities focusing on the needs of children and the elderly is likely to prove particularly challenging. Within quite specific areas it will be necessary to procure larger multi-function properties, without which none of the benefits of the merger can be achieved.

Action

Property records designed for management purposes may be useful for other purposes. One retail chain worked out which of its shops would be flooded if the threatened Thames Surge should occur before the barrier was completed. Hopefully, their plan of action was stored on a shelf above high water mark.

The procedures for the management of a property are closely related to those of the accounts department. Unless rent is received and outgoings paid at the required dates and in the correct manner, the work of the lawyer and the surveyor have been to no avail, since the property has failed to perform its function as a secure investment. In order to obtain that income the landlord will have to take action at key dates and in accordance with the terms of the lease, generally supervise the physical condition of the property and the use made of it by the tenant.

The database permits the surveyor to be advised of key dates when action needs to be taken, the process being:

- extraction or determination of key dates
- selection of appropriate lead time
- entry in perpetual diary
- monitoring of diary
- notification to surveyor
- action taken and amendment of diary
- monitoring by principal.

Whether the clerical procedures involved are the responsibility of a records section or the accounts office or individual surveyors depends upon the type of office and many other factors.

The larger the portfolio, the larger the task. Key dates can arise in connection with:

- statutes
- notices under the Landlord and Tenant Acts
- health, safety, disability and environmental provisions
- structure maintenance schedules and inspection with procedures to enforce covenants
- portfolio valuation dates
- lease rent review procedures: demands and remainders, options
- fire insurance inspection and valuation dates.

In addition, to action under the lease and the exercise of the client's rights are provided for by statute law, the estate manager has to consider broader issues. The performance of individual properties within the portfolio will be viewed in different ways by owner-occupier, tenant or landlord. Opportunities for positive management must be identified, analysed and recommendations made to clients. The portfolio as a whole will be reviewed from time to time and, having regard to the weight attached to various criteria, decisions taken on its management over different time horizons.

Accounts

The account procedures used to give effect to the client's instructions will have developed over a period of years reflecting the requirements of:

- clients
- the firm's business management procedures
- the professional bodies.

The surveyor will need to work closely with the accountants department in respect of:

- form, content and amendment of records
- rent demand and arrears
- reporting forms to clients
- service charges
- VAT
- professional accounting regulations
- firms accounts and fees.

A system should exhibit:

- a facility for rent demands from weekly to annual frequency both in advance and arrear
- the payment of invoices in respect of approved expenditures — the purchase ledger
- the daily recording of cash and bank transactions; tenant's accounts and statements

- arrears notices and monitoring by the surveyor; stop rent procedures (distinguished from zero rent demands); ease of amendment and correction
- financial security
- comprehensive reporting forms to client at appropriate frequency; income statement, expenditure statement for individual properties and a summary sheet, unlet units, service charge analysis
- comprehensive service charge records and apportionment procedures
- ease of operating audit trials
- conformity to professional accounting and surveying standards.

A typical route through which an expense incurred in relation to a client's investment property will pass is as follows:

- invoice received in accordance with contract or instructions
- invoice recorded, considered, approved and signed by management surveyor, passed to accounts department
- client's (landlord's) balance checked to ensure there are sufficient funds for payment — if there is a regular problem, revise arrangements with client
- cheque drawn and entered in payments cash book (computerised or manual)
- client's management ledger debited
- invoice recorded against the property and apportioned amount posted to individual tenants' service charge account, to be demanded in accordance with the lease provisions (in arrear or in advance)
- tenant demand procedures; actual accrual, quarterly in arrear, quarterly in advance
- payment by tenant entered in receipts book and credited to client's management account
- end of year, balancing of advance payments — review of service charge year by management surveyor, prior to year end to establish budget for next year

The need for rent recovery procedures is an inevitable consequence of the service of rent demands. Typical practice consists of the following:

- normal rent demand three to four weeks before quarter day
- final demand reminder on quarter day

- one phone call one week later backed up with arrear letter
- week two arrears letter to be sent by recorded delivery:

> Our records show that the rent in respect of the above property in the sum of £x which was due to our client on the "quarter day" has not been paid despite two applications and a telephone call.
>
> Our clients have instructed us to advise you that if settlement has not been received at this office on or before "date" they reserve the right to take any action open to them to make recovery, including the use of certificated bailiff, without giving further notice, and recovering any costs of so doing from yourselves; should our clients decide to instruct a bailiff then you will be responsible not only for the statutory levy fees, but also any additional commission paid by our clients to the bailiff.

- instruct bailiffs who may recover rent by distraint and if so described (as additional rent) in the lease, service charge
- forfeiture of lease for not payment. Early action by bailiffs; but not always appropriate. Once possession obtained, landlord liable for void occupational costs (rates, insurance, service charge). In a poor letting market it may be better to sue for arrears through court rather than take possession.

Service charges

Service charges must first of all be fully in accordance with the detailed provisions of the lease regarding the service involved, the basis of apportionment, and the frequency of the account.

Problems arise which are not covered by the lease, such as the telephone accounts strike of 1979 where services were provided but accounts were not rendered for up to nine months and then a non standard pattern for a further six months. In equity the service charges should be on a full accruals basis whilst accounts departments tend to prefer to work on a cash basis.

Earlier reference has been made to problems arising where a rating assessment basis is used for apportionment and the effective date varies.

There needs to be a close strict relationship between the accounts department and the surveyor in the supervision of contractors and the monitoring approving and allocating of suppliers' invoices. Once the parameters of the expenditure have been set to a budget, the purchase of services can proceed, albeit with due regard to common sense, case law, and statutory requirements, as well as the all important lease provisions.

Most modern leases do not impose too many restrictions on the manner in which services may be obtained by the landlord. However, it obviously makes good sense when setting up service contracts to obtain competitive quotations from reliable suppliers. This does not necessarily mean selecting the cheapest quote each time, but rather one which will provide the best value for money, provided there is not too wide a discrepancy between the chosen contractor's price and the lowest price offered.

As part of the positive management of the client's portfolio the managing agents should carry out annual analysis of the cost of the provision of specific services. The accountants will need the surveyor's help in determining the most suitable unit for comparison and his opinion as to the relative quality of service provided having regard to the character of different buildings. Long term budgeting eg three to five years ahead, and planned maintenance, will help landlords identify cash flow and problem years. It will also be appreciated by tenants and help avoid acrimonious disputes when large items of expenditure arise.

Specific lease terms must be checked and adhered to, but irrespective of such provisions, all costs incurred must be of a reasonable nature and have been commissioned in a reasonable manner. The age-old question of what constitutes "reasonableness" arises and will vary according to each individual circumstance. There has been case law on the subject, mostly in the context of residential service charges, although the principles can apply to commercial properties. *Finchbourne Ltd* v *Rodrigues* 1976, is perhaps a benchmark case. It was held that not only did the landlords not have the unfettered right to adopt the very highest standard of maintenance and recharge the tenant but it was to be implied that such costs should be "fair and reasonable".

Subsequent cases, such as *Firstcross Ltd* v *Teasdale*, 1982 and *Gleniffer Finance Corporation Ltd* v *Banner Wood and Products Ltd*, 1978, have upheld this principle. Conversely, in *Duke of Westminster* v *Guild*, 1983, a commercial property case, the Court of Appeal advised that implied terms should not necessarily be inferred. This was effectively a decision on repairs and it was held that even though the tenant covenanted to pay the cost incurred by the landlord in maintaining a drain under a private road, it was not implied that the landlord had an obligation to maintain the drain. Irrespective of court cases, it would still be good practice to apply the "fair and reasonable" principle.

In 2005 Jones Lang LaSalle's "Oscar" survey reported the first fall in office service charges for five years:

	£ sq ft *air conditioned*	£ sq ft *non air conditioned*
1994	£4.81	£3.67
1995	£4.46	£3.03
1996	£4.15	£2.97
1997	£4.33	£3.02
1998	£4.49	£3.16
1999	£4.70	£3.29
2000	£4.95	£3.43
2001	£5.16	£3.65
2002	£5.96	£3.95
2003	£6.19	£3.59

The costs are further analysed as to elements:

- energy
- security
- heating/air conditioning
- cleaning
- repairs
- staff
- lifts
- sundries
- water.

Over the 10 year period, cleaning of air conditioned buildings showed the highest increase and repairs of non air conditioned the highest increase. Energy and water costs decreased for both types of buildings.

Geographical location showed:

City	£6.27 sq ft
West End	£5.86 sq ft
Greater London	£4.52 sq ft
South	£3.84 sq ft
North	£3.07 sq ft
Scotland	£3.71 sq ft

Security in the City cost more than four times that in Scotland, reflecting higher threat levels, company requirements and staff costs related to 24 hour services.

Client's money and members' accounts regulations

Chartered surveyors, by their very nature, will handle or hold money belonging to other people. The RICS' bylaw 24(8) states that:

> Subject to the Regulations every Member shall:
> (a) keep in one or more bank accounts separate from his own, his firm's or his company's bank account (as the case may be) any client's money held by or entrusted to him, his firm or his company in any capacity other than that of beneficial owner;
> (b) account at the due time for all moneys held, paid or received on behalf of or from any person (whether a client or not) entitled to such account and whether or not after the taking of such account any payment is due to such person; and
> (c) keep such accounting records as are specified in the Regulations and maintain them in accordance with the Regulations.

On 1 January 1994 new Members Accounts Regulations came into force replacing the 1977 rules. These 15 regulations and associated notes ensure that monies attributed to a chartered surveyor are:

- paid into a separate designated account
- properly recorded in the accounts maintained by the member and
- properly monitored.

It is important that the regulations are read and understood in full by members before receiving clients' money. Members should seek processional advice if in any doubt about their position.

The most important sections of the regulations are set out below.

Regulation 1 — Definitions

- Member: a fellow or member of the RICS in sole or partnership practice.
- Client: any person or body for whom the member acts or any person or body on whose behalf the member holds or receive client's money.

- Connected person: a relation to a member. Money received from such persons is not client money.
- Client account: a current account or deposit account held at a bank or building society into which client's money is paid.
- Discrete client account: means a bank account which exclusively belongs to one client.
- Client's money: all money held or received by a member which is under his exclusive control. It does not include money which the member is beneficially entitled or a float put into a client's account by a member.

Regulation 3 explores the term exclusive control. Any bank or building society account over which a member does not exercise sole control is excluded from the regulations in their entirety. If a member believes he holds money without exclusive control then he should inform the client making it clear it is not covered by RICS regulations or the Client's Money Protection Scheme.

Regulation 4 number, name and nature of the client account. Every member shall maintain at least one client's account with the word client clearly stated in the title. A Discrete Client Account should include the client's name in the title. Balances due to clients should be immediately identifiable and available for repayment. The regulations do not insist on client accounts being interest bearing. However, unless written agreement is received from the client, any interest received belongs to the client or clients. Where non-interest bearing accounts are operated the clients should be notified in writing accordingly.

Regulation 5 deals with payments into the client account. No funds other than client's money should be paid in.

Regulation 7 controls the withdrawal of funds from the client account. A member may not withdraw money from a client account which exceeds the total amount held on behalf of that client. Any money held in a client account owed to the member should be withdrawn immediately.

Regulation 8 qualifies regulation 7 in respect of withdrawals from discrete client accounts. With written instruction from a client it is permissible for such accounts to be "overdrawn", so long as any interest or charges are debited solely to that discrete account.

Regulation 10 — Signatories — Non-members who are signatories to a client account must be covered by a fidelity guarantee in the member's professional indemnity insurance policy.

Regulation 11 — Records
Every member receiving client money should keep accounts to show:
(a) all client money received, held and paid out, or any other money dealt with through the client account, separately for each client
(b) separate current balances for each client.

There should be a separate client cash book, which should be recorded with bank statements and client ledger at least once every 14 weeks.

Regulation 12 — Monitoring
Once every year it is the member's responsibility to prepare and send to the Institution:

(a) a certificate that he did or did not hold or receive client's money and
(b) if he did, an accountant's report and certificate signed by an accountant recognised by the institution and who is not an employee of the member.

Members need to have regard to the Money Laundering Regulations 2003 and the Proceeds of Crime Act 2002.

The client

The general framework within which:

- the client, whether public or private, instructs the surveyor
- the surveyor, whether in-house or consultant, undertakes the task
- the surveyor reports to the client
- the client responds to the report, and his expectations as to the competence of the surveyor

is subject to statutory, contractual and professional standards.

The client, seeking property management services, requires the exercise of professional expertise that lies between that of:

- a valuation report which builds towards the pinnacle of a specific figure and
- a physical report which considers each element in turn where each part is of equal importance.

Property management is essentially tactical, converting policy into the most appropriate action and monitoring its implementation. Over the last few years there has been an increasing recognition of its strategic role. The latter requires a comprehensive understanding of the client's business, whether that be delivering healthcare in a large concentrated urban acute health trust, or a nationally dispersed retail banking portfolio of 2000 branches.

There is a growing recognition on large, particularly multi-use development schemes, to appoint the property manager as part of the development team with the brief of ensuring design meets tenant requirements, minimising operating costs and co-ordinating service providers. As a result there should be much more confidence in the services and their costs.

In the surveyor's action on behalf of his landlord client, the tenant is the landlord's client. In an RICS survey of tenant satisfaction in 2005, the results were not encouraging, with scores out of 100 for:

Landlord/tenant communication	43
Contract detail	42
Problem resolution	47
Lease flexibility	33
Value for money	46
Location of premises	63
Standard of premises	58
Overall tenant satisfaction	39

Of course this is highly subjective and gives an illusory sense of accuracy.

The survey was based on interviews with 66 tenants, more than half of which pay rents of £500,000 or more. There is no doubt that better communication would also improve the other scores. Some landlords, particularly those with large distribution portfolios, have recognised that just in time management means timely space provision for customers across a network of depots, which can be best delivered by service level agreements rather than leases.

Instructions and fees

The instructions given by a client and accepted by the surveyor on behalf of his firm create a legal contract between the parties. In the past, instructions have not been approached with the priority they now receive, partly due to the Unfair Contracts Terms Act 1977. Also, clients are becoming increasingly sophisticated, aware of the subtle relationships between instructions, fees, caveats and standards of competence. Some tasks carried out by the profession are well known, easily described, and the results specific and immediate. In these circumstances instructions and their acceptance should present few problems. However, the more specialist the property, the less routine the task, the more unstable the market, the longer the life of the projects then, the greater is the degree of professional expertise required. This has implications as to the calibre and experience of the professional staff involved and the appropriate basis for fees. In these circumstances the original indication of instructions from the client may need refinement and this is often supported with preliminary advice. Such discussions are rather difficult since, at this stage, fee negotiation will also be taking place.

The levels are which management services may be provided are illustrated by the following from the management services brochure of a major firm:

> The combination of clients' resources and needs, interacting with the different types of property and terms under which they are occupied, produces a requirement for the provision of different levels of professional management.
>
> These include:
>
> *Watching brief*: the minimum level of management appropriate to vacant property awaiting redevelopment, where we respond only to statutory notices or similar requirements
>
> *Care and maintenance*: appropriate for buildings which are vacant for a variety of operational or investment purposes
>
> Standard management: management of an individual building with normal lease covenants and service contracts, rent collection, service charge and accounting work
>
> *Portfolio management*: this implies a comprehensive application of management skills across a number of properties with close liaison with the client leading to the implementation of a plan for the portfolio. Typically, this will include a diary of: rent review, insurance, inspection, schedules, action. It is here that we are most able to demonstrate positive management
>
> *Investment performance management*: this is a logical development of portfolio management and utilises our capacity for the preparation of analytical reports on the performance of portfolios on both sector and geographical bases. These are then related to appropriate indices enabling recommendations to be made on individual property performance.

Having identified the level of management, the duties and frequency at which different activities should be undertaken can be established. The previously mentioned document refers to:

> The continuing nature of management means that cycles of activity can be identified and this enables a periodic management system to be developed.

Regular action will include:

- *daily*: response to tenants' enquiries and emergency action in the case of building or services failure; the "help desk"
- *monthly*: monitoring of the quality and efficiency of those services provided
- *quarterly*: the issue of rent and service charge demands action on non-payment and reporting and accounting to the client
- *annual:* cycles of inspection for redecoration, maintenance and repair, together with forward budgeting
- *every four or five years*: a major review of policy and implementation, rent review and lease renewals.

The result of agreeing the level and frequency of management activity can be a management brief. This is illustrated by the *Duties of Managing Agents of Commercial Property*, published by RICS in 1985.

> For buildings let in single occupation on a full repairing and insuring lease, for a standard management fee the managing agent should:
>
> Submit demands for and collect rents, insurance licence fees and other charges from lessees on a quarterly basis and forward monies at agreed intervals
> Make payments to superior lessors, professional advisers, insurers or others, subject to RICS accounts regulations
> Check that tenant has current building and other insurances, and administer other insurances, such as public liability cover, on behalf of the landlord
> Maintain separate clients' accounts and ledgers and prepare and submit statements and accounts with balances and supporting invoices at agreed intervals, normally quarterly. Inspect property periodically to ensure proper occupation and to check compliance with lease as to decoration and repairs
> Deal with routine management enquiries from tenants
> Keep records in relation to tenancy and other relevant matters relating to the property.
> The agent should also advise on obligations to serve statutory or contractual notices, such as rent reviews and lease expiries and options, before the

notices fall due. He should instruct solicitors or certified bailiffs on unpaid rent, charges or other matters agreed with the client, including late payment of monies due under the terms of the lease, he will attend court regarding arrears of rent and other monies.

Consider applications by tenants for alterations
Offer professional service in connection with applications for the lessor's consent to assignments, sublettings and changes of use
Advise on management policy
Provide inspection reports on general conditions to clients
Liaise with client's solicitors in relation to leases and other relevant documents

For buildings in multiple occupation with the landlord responsible for the provision of services, the managing agent should:

Submit demands and collect payments for service charges — quarterly — or to meet lease requirements, deal with annual reconciliation of these charges and administer the charges for residential tenancies
Pay suppliers, contractors, staff or others providing services. Administer building and other insurances, including plant and equipment, employer's liability and public liability cover. Produce and circulate service charge accounts, including balancing charges ore refunds and, where necessary, supply information to tenants and others unless leases require certification by independent auditors
Where necessary, produce annual estimates of future expenditure for service charge purposes and reserves, administer funds and provide information to auditors
When instructed by client, engage and supervise staff, detail their duties and pay wages and attendant expenses. Arrange and supervise standard maintenance contracts for all plant and equipment, general cleaning, security, landscape maintenance and others
Inspect property periodically to check day to day running and operation
Deal with routine management enquiries from tenants about services provided
Deal with day to day repairs to structure, plant, fixture and fittings and equipment, and with emergency repairs. Keep records on tenancies and other matters relevant to property
For shopping centres, maintain overall responsibility for promotions and administer promotions account. Check compliance with terms of lease and statutory requirements
Arrange for fire alarm systems to be tested and for regular fire practices and, where appropriate, arrange annual boiler and lift inspections
Where appropriate, arrange load tests on cradles, safety eyebolts or other lifting tackle and annual changeover on sprinkler systems from water to air and vice versa. Submit applications for fire certificates

Where appropriate obtain petroleum spirit licences for internal car parking areas
Maintain on premises accident book, copies of fire certificate and other statutory reports and fire alarm log book, and administer matters arising
Deal with day to day matters concerning safety and welfare of people within property arising from statute or other regulations.

Managing agents may agree other duties with clients which fall outside the scope of the standard fee. The agent may:

Prepare specifications, obtain competitive tenders and supervise works of substantial nature other than day to day repairs
Give evidence in court on recovery of rent or other charges from tenants, or compliance with lease covenants
Advise on rating, planning, grant applications, or insurance claims
Prepare replacement cost valuations for fire insurance purposes
Prepare schedules of dilapidations or condition, negotiate claim settlements or supervise repairs
Negotiate rent reviews and submit evidence to arbitration or independent expert
Effect lettings or renewals of leases or other options and attend court hearing sin connection with lease renewals
Prepare and advise on valuations for insurance, capital or rental purposes or capital allowances
Provide copies of leases, policies, receipts and other documents or additional copies of current or past accounts
Deal with matters arising in connection with adjoining properties or owners
Advise on renewal or modification or restrictive covenants
For residential tenancies, offer traditional professional services on matters arising under Rent or Housing Acts.

In 1994 these were followed by *Advisory Guidelines for Commercial Property Management*, set out in Appendix F. These provide an equitable basis for the relationship of landlord and tenant achieved by the conduct of the managing agent in the operation of lease covenants. Management brings with it review and renewal, insurance inspections and valuations, maintenance and refurbishment, acquisition and disposals, rating, and other specialist services, available from a firm able to provide a comprehensive service, each of which generates its own fees. The certainty and regularity of management work means that it can be staffed efficiently, providing a stable element of the firm's fee income, and since part of the fees, a percentage of service charge is cost-orientated and part a percentage of income received is value-orientated, there is a further in built stability.

In *Concorde Graphics Ltd* v *Andromeda Investments SA*, 1982, there was a service charge with a schedule of items which did not specifically include either the managing agents fees or a basis upon which they should be calculated. However, the landlords employed the firm of Wimbourne Martin French & Co who, in their accounts to tenants, included a charge for management based on 10% of the expenditure incurred by the landlord.

New landlords appointed new agents, Grant & Partners. They adopted a new basis for their managing agents' remuneration, namely 7.5% of the rental income. The tenant, on comparing the old and new management charge, noted the new charge was almost three times higher, and that the management fee element on the new basis which had been agreed with the landlord, was nearly five times higher. It was held that the managing agents were not suitably disinterested to determine the dispute as landlord's surveyor.

The fees charged for residential management are illustrated in the case of *Parkside Knightsbridge Ltd* v *Horwitz*, 1983. The management of large portfolios are frequently subject to tender, possibly as part of an overarching framework agreement. If in the public sector then possibly under the auspices of the Office of Government Commerce, their financial worth may fall within the OJEC criteria where a particularly formal process must be adopted.

The more rigorous and comprehensive the tender specification and the more effort the client puts in to creating a competent and compliant shortlist, the better will be the outcome. A three year appointment with an expectation of five years is normal, in order for both parties to benefit from working together.

While pricing is the patent outcome, the latent inputs are the decisions on how the consultant deploys the resources:

- A dedicated or generalist term?
- A national or regional response?
- Horizontal or vertical management?

How is the service to be organised so that the engagement partner has the necessary command and hence, confidence, of the client as he internally procures services from sectors and offices at prices that individually are unattractive?

From time to time potential conflicts of interest arise. Generally, this will be because of some existing client relationship in the property though there may be other reasons concerned with personal or family involvement. Once the client has been informed, he may well be happy for the firm to continue to act. It can sometimes happen that, with the consent of both parties, a firm can act for both parties simultaneously, for example, two partners could hold personal appointments to particular bodies. Within the robustness of a large firm and with the clear consent of both parties in some formal exchange of letters, a quite adequate professional relationship can exist.

Inspections

Inspections are a key requirement for a good records system, which is essential for sound management. There are many reasons for property inspection — the majority are related to the creation, enforcement or transfer of rights in the property itself or transactions in financial assets which offer interests in the property. The remainder of inspections are mainly in connection with the physical fabric of the building, though very often they will be as a result of the enforcement of rights and duties between the landlord and tenant. Property inspection is not just a physical matter, it provides an opportunity to make personal contact with the occupier, be he tenant or owner, with whom the property manager will be in regular communication.

No estate manager likes to take decisions in respect of property he has not personally inspected. In large organisations this is often inevitable and in these circumstances the estate manager will need to have confidence in both the ability of the surveyor and the comprehensiveness of procedures to ensure that the right information is available within the management organisation. The link between the formulation of policy and its implementation in respect of individual properties needs to be more than a matter of personalities; they come and go, suffer illness and are subject to the normal human failings. In large organisations, to provide job satisfaction and so motivate professional staff to give of their best, the executive manager and the property surveyor need to have an effective form of communication, in both directions. The larger the estates organisation, particularly in the public sector, the more likely it is that a technician grade of property referencer is responsible for the detailed referencing and recording of property information and its subsequent storage and access. Some very good surveyors take a very bad reference, concerned only with the information necessary to solve the particular instructions rather than the creation of a good record of all aspects of the property which will serve to provide useful information on several other occasions in the future. At the same time technicians can only be effective with the benefit of detailed professional supervision. Too little use is still made of photographs in recording the character of buildings, which can seldom be achieved by the most copious notes.

The first requirement before a detailed inspection is a clear brief as to its purpose, and existing records should be carefully checked. In the public service or commerce and industry, that is to say, single client in-house estates departments, this is very much easier than in private practice where the multiplicity of clients means that the filing systems tend to be client-based rather than property-based.

It is important to identify any previous records in connection with a property upon which the firm has been newly instructed. In the past the property may have been disposed of by an existing client, used as a comparable, investigated on behalf of another client or in one of several other ways already be well known to one or more individual; this must not create a conflict of interest.

The result is unnecessary time spent on new referencing and lack of information on the history of the property which may be difficult or impossible to obtain at the present time. Accurate measurement based upon the RICS Code of Measuring Practice is undoubtedly one of the requirements for a successful negotiation. To enter a negotiation knowing that all your measurements are accurate to the nearest 3ins (or 7.5cm) can be used to advantage. If differences occur and the other surveyor is not prepared to accept your figures, a joint inspection can have a rather chastening effect, particularly if you indicate that you regard agreed areas as important for your proof of evidence. Better still, the presentation of a complete typed schedule of areas at the very first meeting can impress upon the other party your determination to pursue your client's interests to the full. On the other hand, it is easy to spend far too much of the site visit carrying out meticulous measurement and failing to recognise qualitative and subjective factors which fundamentally affect the valuations and the management advice arising from a report. The pursuit of accuracy of 0.01% in measurements used in calculations, when accuracy of many of the other factors influencing the decision are around say 5%, introduces a false sense of security into what is not a science, perhaps poorly described as an art and more akin to a craft. The wide availability of CAD drawings and the ease with which key dimensions may be measured electronically, means inspection with CAD drawings and check measurement is more likely to produce accurate areas than comprehensive on-site measurement, particularly in difficult buildings.

A surveyor will determine the most appropriate procedure for an inspection. In large complexes a preliminary walk about with the benefit of the opinion of site management can be followed with some time with the site engineer who can usually find copies of plans if these were not already available. The detailed referencing technique is a matter of personal preference. There are some benefits in taking the largest overall measurement possible and then taking out. However, those with experience of steel and paper mills, lard factories and killing lines at meat pie factories, tend to adopt a more pragmatic view.

Many of the factors upon which advice is based cannot be measured but some can be monitored on a regular basis. Regular inspections and prompt action to remedy defects can have a positive effect not just on the fabric of the building but on the attitude of the individual tenant involved, as well as protecting the client from various statutory liabilities. The overall integrity of the portfolio depends upon many factors, not least of which is the opinion held by the occupiers of the quality of the management.

For some properties, mainly those valued having regard to trading potential, many other features need to be considered during inspection. These relate to the capacity of the property; hotel, theatre, petrol station, licensed, leisure, waste; to trade compared with the actual trade in physical and financial terms.

Reports

Skill in report writing can only be achieved after years of experience, it is salutary for a surveyor to study a report he wrote several years ago and realise how his style has developed. There is no doubt that a poorly constructed or badly phrased report detracts from the client's appreciation of, and confidence in, the recommendations made in a report. The ability to draft a good report is fundamental to the effective exercise of the surveyor's expertise, it is the main form of communication with the client and, apart from the property itself, the permanent record of fact and opinion upon which further instructions may be taken. The several different types of reports illustrate the range of work associated with the management of individual properties or the estate as a whole.

A valuation report, though requiring adequate descriptive material and some consideration of either the statutory or market setting, tends to turn upon the figure that appears in the last paragraph; there is a measurable financial result and this may be tested in the market place.

A report dealing with the physical condition of a property, such as a structural survey, does not build up so much to a focus, but presents a comprehensive view. A management report lies somewhere between these two types of report. Property management aspects will be considered, as will the potential of the physical structure to meet the needs of occupiers and its suitability for other uses. Several alternative courses of action are usually identified, the report concluding with the reasons for selecting the most appropriate action and identifying possible consequences.

The report should result in the reader arriving inevitably at the same conclusion as the writer, with a degree of appreciation appropriate to his technical knowledge. The surveyor must have a clear understanding of his client's perception of the issues which are being considered. There is no blueprint that can be used, but the report should be confined to meeting the instructions.

An initial meeting with the client is helpful in establishing communication and clarifying the scope of instructions far more effectively than correspondence, though a written record of the discussions should be confirmed with the client as soon as possible. Fact and opinion should be sufficiently separate for the client to appreciate the difference, but linked so that opinion can be analysed, perhaps in the concluding paragraphs. The report should stand as a comprehensive document, but this does not mean it must be either lengthy or complex. A short report with several appendices or schedules of fact may be as effective as an apparently longer report.

At various points in the report it will be necessary to digress from the main flow to define or illustrate technical terminology. This is worth doing well once, early in the report for each term, rather than covering somewhat inadequately in a different form of words on each occasion. A covering letter attached to the report enables any necessary personal communication to be achieved. The report itself should be in an impersonal style and verbosity can often be avoided by making sentences short, which prevents the convoluted sentence within which the extravagant use of English seems to flourish: this sentence is showing symptoms of verbosity. Lastly, it concentrates the mind most effectively to say to yourself, "How would this sound in court in a year's time, if I were being sued for professional negligence?"

Competence and compliance

The action of a surveyor in managing an estate or individual property on behalf of his client can be subject to criticism on account of several types of misconduct arising from:

- criminal law — both general and specific
- Estates Agent Act 1979
- Consumer Protection Act 1987
- Misrepresentation Act 1979
- Property Misdescription Act 1991.

In respect of the latter, a statutory instrument prescribes 33 matters which are an offence to misdescribe, the avoidance of which requires comprehensive and effective office systems.

- extensive property statutes
- statutory instruments
- circulars
- tort, including negligence
- professional codes
- stock exchange
- RICS — account regulations, valuation standards, model clauses
- clients instructions — confidentiality
- professional practice generally
- firm's code and quality assurance procedures
- personal conduct.

Having reached the end of the hierarchy, is there anything left to say? We are all influenced by our own experience and that of role models throughout our professional life. Perhaps the final criteria is to consider how one of those role models would have approached the task, analysed the problem or reviewed the work with the client. Debate with colleagues, even heated discussion remains the best way of testing and improving competence.

While there is no natural symmetry between something going wrong and negligence, there is no doubt that clients have regard for the advisor who identifies an error and offers options to correct it, and agree the way forward (on an infrequent basis).

The careful acceptance of instructions and the selection of appropriate caveats should ensure that the surveyor enters into a contract that he is potentially capable of carrying out on behalf of his client. However, problems do occur under the head of negligence. In contract, a claim is time-barred after six years, where as a claim in negligence extends to six years after the date when the loss occurs, which may be several years after the completion of the job. This means that insurers are faced with a long tail before any given year of account can be finalised. A retired partner would be wise to ensure that adequate professional negligence cover remains in force to cover claims arising from when he was in practice.

The standard of care required by a professional man was considered in the case of *Bolam* v *Friern Hospital Management Co*, 1957, where it was stated that

> the test is that of the ordinary skilled man exercising and professing to have that special skill. A man need not possess the highest expert skill: it is well established law that it is sufficient if he exercises skill of an ordinary competent man exercising that particular art.

Some general principles have emerged from various cases:

- the standard is not absolute
- the onus is on the plaintiff to prove the defects of conduct
- the ordinary general practitioner is not judged by the standards of the acknowledged expert
- if a surveyor holds himself out to be a specialist, that is his standard
- the inexperienced are to be judged by the same standard as the experienced
- he following of standard practices will make the burden of proving negligence more difficult
- an advisor must take care in recommending a specialist
- a reduced fee does not affect the standard.

The work of the surveyor in property management can best be illustrated by the following cases. The taking of initiatives under leases is perhaps the most practical example of positive management. The negligence implications of the decision in *United Scientific Instruments Holdings Ltd* v *Burnley Borough Council*, 1977, that time was not of the essence, caused many a sigh of relief in lawyers' and surveyors' offices. Though where time is explicitly of the essence, as in the case of options, the financial consequences can be severe. In *Social Workers Pension Fund* v *Wood Nash & Winters*, 1983, there was a 21 year lease with an option to break in favour of both parties in the 15th year. The solicitors involved failed to advise the fund as to the exercise of the option with the result that whereas market rent could have been obtained, the tenants continued in occupation at an historic rent for a further six years. Damages of £95,000 were awarded.

Action under leases is substantially influenced by the 1954 Act: in *Howard* v *Woodman Matthews*, 1983, solicitors for the tenants failed to make an application to the County Court for a new tenancy. As a result, the tenant had to accept renewal on the landlord's standard lease for only three years. Damages were awarded.

One question frequently raised is how much a non legal professional should know of the law in his chosen specialism. *BL Holding Ltd* v *Robert J Wood & Partners*, 1979, provides some useful guidance. An architect was instructed to design a building within the 10,000 sq ft exemption from the then Office Development Permits. The defendants were surprised to learn from the planning authority that the basement car park and caretaker's flat could be ignored in the calculations but the building was designed and built on this basis. Sadly, the building proved impossible to let since prospective tenants were told that the planning permission was invalid, as the necessary permit for a building of over 10,000 sq ft had not been obtained. The High Court held the architects negligent as they should have pursued their doubts and advised their clients to obtain expert legal advice. The Court of Appeal, 1979, allowed an appeal on the grounds that it was a difficult area of law, likely to confuse other architects and the plaintiff, advised by chartered surveyors and experienced property developers, had independently taken the same view as the defendants.

In *PK Finans International (UK) Ltd* v *Andrew Downs & Co*, 1992, the valuer accepted two documents from the owner: a planning consent for nursing home and a schedule referring to change of use for residential nursing home and 30 warden-assisted residential units. He valued on this basis and made no reference to need for verification. Although this was contrary to a RICS Guidance Note, the valuer was held not to be negligent (the author does not agree).

The taking up of references is perhaps regarded as something of a mere formality; this was considered in *Nahhas* v *Pier House (Cheyne Walk) Management Ltd*, 1984: the defendants were Harold Williams & Partners who were engaged as managing agents. The flats were expensive: in 1984 a one-bedroomed flat cost £55,000 and in 1980, a typical service charge was £2,000 per annum. A complex series of events including the occupiers' illness, sets of keys, replacement locks etc, resulted in the porters holding the keys to the flat and later it was found that jewellery of considerable value was missing from the flat. One of the porters pleaded guilty and it became clear his profession was that of thief, having served 11 sentences since 1966 with 33 convictions. The court held that the key control system was quite satisfactory but there had been negligence in the recruitment of the porter and particularly in the checking of references since the porters were, in effect, in a similar position to security guards.

Following on from the breach of the duty of care, the measure of damages and the mitigation of loss needs to be considered. Perhaps the most interesting cases for the property manager is *Teasdale* v *Williams*, 1984. The damages in a negligence case turned on the timing in the s 24a notice and its relationship to a rising market. The landlord's solicitors had served an interim rent notice effective from June 1977; unfortunately, despite several requests, the tenant's solicitor failed to advise the tenant's surveyor of this. The latter continued to defer negotiations for a new lease, thinking his client continued to pay the historic rent of £4,500.

If a new five year lease had been agreed with effect from June 1977, when the plaintiff's original term expired, it was estimated the rent would have been £20,000. The rent was not agreed until June 1979 at £26,500 for a seven year lease. The landlord's surveyor was able to negotiate a reduction for the first two years to £19,000. It was held that the consequences of the tenant's solicitor's negligence was:

A loss of 3 years × (£26,500 – £20,000)	£19,500
A gain of 2 years × (£20,000 – £19,000)	£2,000
	£17,500

The defendant had argued that if a rent had been negotiated as at 1977 on a five year term then, upon renewal of that lease in 1982, there would have been a saving compared with the actual events that occurred, since the new hypothetical lease in 1982 would have been at £40,000 pa, whereas for the first two years the tenant was actually paying £36,500 pa. However, the court held that this possible temporary rental advantage could not be isolated.

It must be acknowledged by even the most experienced and respected senior partner that, despite his own knowledge, the firm may through the excesses, inadequacy or recklessness of one individual — or misfortune — be faced with a claim for negligence. The firm will then read its professional negligence policy with deepening interest.

Large firms typically hold cover for over £100million, with substantial excesses for claims at £50,000 or £75,000 each. Premiums have risen significantly from around 2% of turnover in the late 1990s, to 3%, to 4% in 2005, hence a bad claims record impacts on profit.

In common with other professions, particularly accountants, lower limits should be agreed on individual instructions and best practice applied ruthlessly across the business. The apparent attraction of limited liability partnerships deals only with the doomsday scenario, but they are being adopted by all large partnerships.

It may be thought that many of the preceding matters have little or no relevance to the surveyor in the public services or commerce and industry. This is not so; the standards of competence to be expected are the same for all those who claim to exercise a particular professional skill. It is true that responsibility may be less personal and retribution more diffuse in character, bringing the organisation rather than the individual into disrepute. In local government, changes in the political control of the council can bring about policy changes which result in the acquisition of disposal of land in a manner quite outside the principles of prudent estate management.

The best professional negligence policy is the encouragement of good practice at all levels within the firm. Apart from a sound knowledge of the law, a positive approach to avoid claims and hence keep the premium low, would include the following:

- written instructions, acknowledgement and written clarification of any uncertainty, including agreed file notes on meetings
- standard procedures for standard types of job, using job cards or proformas and progress reports
- the efficient handling of correspondence received is much enhanced by the use of a bring-forward procedure to prevent letters being permanently filed until all their contents have received attention
- supervision by the principals in some convenient form, either at certain stages during jobs or periodic briefings. In larger practices, direct supervision by the principals can sometimes be a problem
- regular communications with the client by means of interim reports. Clients may assume, correctly in some cases, that no information from the firm means no action is occurring
- rigorous diary and accounting procedures which provide an independent form of monitoring the work of junior staff
- the ability to recognise the circumstances in which the client should be recommended to seek specialist advice.

Trust must still be placed in the competence, integrity and sense of professional enquiry of staff. This is undoubtedly one of the attractions of the profession, the ability to use independent thought to resolve client's problems and, in so doing, develop one's own personal skill in the practical application of professional expertise. It brings with it a heavy responsibility and the surveyor must appreciate this early in his career. Equally, the employer should provide an environment in which a high level of professional competence is encouraged and demonstrated by experience practitioners.

Quality assurance

In 1992 the RICS published *Quality Management for Valuers*, the objectives of which were to:

- accurately interpret client's instructions
- deliver advice promptly
- continued care of client during the commission
- obtain confirmation that instructions have been fully complied with
- obtain confirmation of acceptance of fee
- periodically monitor the provision of the service
- clearly measure added value to client
- monitor client's requirements and the service provided.

By following a system of management practices, a company can assure good quality and a consistent service. A company that does not employ good management practices has to deal with the consequences of failure or poor quality. Quality "Assurance" (QA) is similar to the principle that prevention is better than cure.

ISO9001 is internationally recognised and certified by independent accredited bodies and maintained by external and internal audit. This provides clients with the confidence that the company's service is continually under review and "assurance" that they will continue to receive a good quality service.

The benefits of a QA system flow from consistency, staff across a network of offices deliver work to agreed standards and clients know the process and have the benefit of standard formats, this requires regular review by internal audit.

This makes staff more effective, agreeing instructions is easier and offers reassurance to the procurement of professional indemnity insurance. If problems occur, whether or not formal complaints, a process exists to review, improve and implement change, focussed on satisfied clients. Satisfied clients return to give the company more business.

The starting point of a QA system is a company's "Quality Policy" or "Mission Statement". This must state the company's objectives with respect to quality and must be approved by top management. Documentation is conventionally organised in a three tier structure.

- *Quality manual*
  This is the top level document which typically contains the mission statement, a company organisation chart, outlines the structure of the QA system, briefly describes the company's processes and how they relate to each other.

- *Procedures*
  Procedures are generally mandatory. Essential activities are described here and there would normally be a separate procedure for each type of surveying practice, twenty or more in a large practice.

- *Guidance notes*
  The third tier contains "best practice" such as guidance manuals, local work instructions and any forms in general use.

A QA system can be based on paper or an electronic system, such as a company's intranet or a bought-in proprietary system. Whichever system is implemented, documentation must be controlled so that only up to date information is used. With an electronic system, a company does not have the complication of withdrawing old procedures issued on paper.

Continual improvement is a benefit to a company and its clients. A company would usually review client feedback (eg complaints and client surveys) and the results of internal and external audits. Data is analysed to identify any common problems and long term trends that need to be addressed.

Many companies have their procedures written by staff that like writing too much. Sometimes this results in long, wordy procedures that are not easy to read or use. Concise procedures containing only essential requirements are more effective and easier to maintain. Professional organisations can often provide guidance on the best way to write procedures. A major pitfall is complacency, this often occurs a few years after implementing a QA system, when staff begin to go back to the way they used to work or the culture fails to deliver. A company must not assume that once the system is documented, it will look after itself. A QA system requires constant review and maintenance to ensure that clients receive the service they require and expect.

Changing technology

The exponential increase in computing power, data storage capabilities and communication technologies in recent years means that both computer systems and the opportunities to exploit them are constantly changing.

The Internet provides ease of access to information from anywhere in the world and email enables quick and easy communication. Mobile phones, laptop computers and hand-held devices are changing working practices in many areas of work, and estate management, where staff can be out of the office or on the road for a significant part of their working day is benefiting considerably from these technologies. The increasing demand for access to accurate and up to date information is now influencing the way data is stored and managed.

Information management

As with any other business, property companies need administrative systems for corporate functions like accounting, human resources and training. The software products used are not necessarily specific to the property market, but are often generic products which can be tailored to meet any property-specific requirements.

This also applies to the client relationship management function, where systems suitable for the service industry can be used, although it is more beneficial if they are integrated with other core systems.

In addition to recording basic company and contact name and addresses, contact management systems enable clients and other contacts to be classified as required. Market sector is an obvious example (eg finance, manufacturing) and also company type (eg bank, pharmaceuticals). It is also important to record the relationship with each company and contact, as they may be clients, potential clients, or those in an intermediary capacity (banks, solicitors etc), and also who knows them. Other details such as their property and/or location specific interests can also be recorded.

Contact classifications can be used for selective targeted mailings to ensure contacts are aware of the services offered, or to inform them of new services. The system will then hold a record of mailings sent to each contact and if required can also store a copy of the letter or brochure mailed.

Most contact management systems also provide event management functionality, which as well as providing the ability to mail invitations, enable a record of replies to be stored so that events can be monitored.

A well-designed system can help to ensure regular communication with clients, enable the monitoring of work with them and keep them updated with the services offered. It can also ensure that the company keeps track of the clients changing requirements. In a multi-disciplined practice it is important that there is corporate management of the multi-layered relationship with each client.

A comprehensive instructions database is central to a firm's instructions management, recording details of the client, type of instruction, property, any other party eg tenant or intermediary. This provides the information required for checking any conflict of interest. The database can also be used to record estimated or agreed fees and anticipated billing dates, which provides for reliable fee forecasting. The progress of instructions of varying types can be monitored including accounting for "work in progress".

Time and expense recording is also becoming more widely used in the property field with surveyors recording time and expenses against individual instructions. The hourly charge-out rate for each surveyor and any agreed budget can be set per instruction and costs can then be monitored. This information can be used to support the fee to the client by a building surveying department for example. A professional department like planning may use it to monitor time spent on an instruction against an agreed budget and a property management department may use it to monitor the cost of servicing specific instructions.

Instructions processing will vary with the type of instruction and the following gives some examples of different types of system, which may be separate or modules of an integrated system.

Commercial and residential agency systems provide the means of recording property requirements and available properties and the facilities for matching them. Many also enable the recording of any deals done and thus provide comparable evidence. Agents are generally more anxious to move on to the next deal rather than record details of the one they have just completed, so it is essential to have triggers which force the recording of deal information as instructions are completed to ensure that this valuable data is collected.

A successful agency system is dependent on structuring the system and defining coding structures to provide the appropriate classification of applicants' requirements and available properties. A retailer may have very precise property and location requirements while an investor may consider multiple property types and be very general about the location but precise about the yield required. Many companies now also publish available properties via the Internet and provide searching facilities to enable users to search for properties on-line. Available properties are either published to a corporate web site or via feeds to other data providers.

Specialist systems are available for project appraisal and development and investment valuation analysis. They can handle single and mixed-use developments projects, overlapping time phases, turnover and stepped rents and provide sensitivity analysis. The source data for analysis is often available in other systems and the ability to import and export data between systems is essential.

Property management and accounting systems enable investors, or managing agents acting on their behalf, to manage their property portfolios. They provide the accounting for rent and service charge collection, together with a property database to enable effective management and reporting. Many property management systems are modular with optional modules for additional functionality like a Helpdesk for tenants or to record planned maintenance programmes. Other systems are focussed on the needs of the occupier rather than the landlord. These systems may include an asset register and provide facilities management.

Graphical Information Systems (GIS) allow properties to be located precisely on a map. Many systems containing address information have links via the postcode to a web-based mapping system and it is also possible for a whole property portfolio to be presented on a map. Other systems can be used to help make informed decisions about existing or potential locations, calculate drive times to specific locations or provide demographic information.

There are large quantities of external data available to support a surveyor's work and the Internet has made access to this information easier than ever before. Whereas much information used to be available only in hard copy or on request, many government departments and quangos now publish very detailed information on their websites. For example:

- National Statistics Online *http://www.statistics.gov.uk/* publishes very detailed figures on every aspect of UK life
- The Land Registry *http://www.landregistry.gov.uk* produces statistics on property prices with data detailed down to postcode level. An online ordering service also gives fast easy access to data that previously required time consuming submission of forms by post
- The Valuation Office *http://www.voa.gov.uk* offers a fully searchable online version of the Rating List free of charge.

There are also many commercial organisations offering property related information services on a subscription basis. Some of these can be installed in-house but increasingly these services also are delivered via secure websites. Services like Focus and EGi have large teams collecting and analysing market data and offer this information via a user-friendly interface with powerful search functions. Once strictly the domain of research departments, many firms now make these services available to all staff. It is relatively simple to retrieve information on property deals, availability, companies' property involvements or requirements.

Firm's management

Whether or not a company has invested in fully integrated data systems, there are technologies and reporting tools available which enable the consolidation and analysis of data from multiple databases in order to provide the global view. This can range from the analysis of fees to identify top clients to the fee income per individual fee earner or profitability of a specific instruction. Reports can be run as required or prompted by pre-defined alerts or triggers.

Instructions can be monitored to establish trends in types of property or instructions. Fee income can be analysed in specific market sectors, types of work or property types.

The profitability of specific types of work can be analysed although, as the major resource of a property company is its staff, it is difficult to monitor the profitability of specific instructions without implementing time-recording systems, standard practise in the legal sector. While this is accepted as essential for some types of work there is understandable resistance in transactional markets and few professional practices have incorporated this fully.

As professional firms increase in size, particularly in multiple locations, it can become increasingly difficult to maintain effective communication. Keeping staff informed of business developments, giving opportunities for cross-selling and also of management and administrative issues is critical. The increasing use of email can help with this but can equally contribute to information overload.

Many larger firms are now deploying intranets as key corporate communication tools. An intranet is effectively a website for internal use only; information can be shared via a user-friendly web interface but is secure within the firm's firewall — a firm wide web. An intranet can act as a central reference point for the firm enabling information and expertise to be easily shared and kept up to date. Typical information might include:

- major client instructions and deals news
- corporate announcements relating to management and strategic issues
- information on company performance
- company policies and procedures
- information to support tenders and pitches
- templates and standard documents
- data sources for sharing
- links to databases and applications
- discussion boards
- contacts database
- technical information and legal updates.

Changing requirements

Many clients now expect access to information held on their behalf, whether by direct access to their data or via a shared extranet. Investors want access to their managing agents' systems to monitor rental income and arrears and occupiers to their property portfolio details. They require live access to their data and use it to monitor the performance of managing agents. Data can "pulled" by the client, ie made available on demand, or "pushed" to the client by pre-agreed triggers. Some investors permit tenants to view their accounts records stored in a property management and accounting system and access is often also provided for building managers.

There is increasing demand for systems to handle pan-European or global portfolios with the requirement for multiple languages, currencies and measurement functionality and also the ability to handle country specific requirements for fiscal and property legislation. Managing agents are also increasingly required to use the client's own property management system.

Home working is becoming more accepted, and surveyors require access to information systems from home or on the road. A significant amount of correspondence with clients and business partners is handled via email, and email systems and diaries need to be accessible at any time. Surveyors also require database systems to integrate with their email system so that alerts can be emailed or a diary reminder of forthcoming events (eg rent reviews) can be recorded. There is also demand for corporate contact databases to integrate with personal email address books.

Data management and security

IT departments have the difficult task of maintaining the security of their systems and network while making data more accessible. Internal networks are protected by firewalls with access via secure gateways for offsite clients, staff and business partners. Extranets can provide shared access to information with third parties. In multi-disciplined practices information may be confidential to one part of the practice and there may be a need to set up "Chinese walls" to restrict access to information.

With much business communication handled electronically, there is a movement away from maintaining hard copy files. Emails are regarded as legal documents and in the event of a dispute a court may request the disclosure of all relevant documents, including email. This means that electronic documents must be stored for six years and documents under seal or deed for up to 12 years, and in the case of FSA-regulated companies, be retrievable within 24 hours.

In the past, property computer systems were often implemented for a specific function, for example a contact management, or agency function, with the resulting duplication and inconsistency of data. Many IT departments are now faced with the task of consolidating these legacy systems and developing corporate integrated systems. There is also a move away from traditional client-server delivery of systems to web-based development, which enables systems to be more easily distributed and accessed.

Property data is difficult to reference and locate precisely. While postcodes and grid referencing can assist with searching and retrieving there is a recognised need for a common standard. The *National Land and Property Gazetteer* (NLPG) *www.nlpg.org.uk* is an initiative of the National Land Information Service (NLIS) to provide unambiguous identification of land and property and hence provide access to the information required for the conveyancing process. Local authorities are the definitive source of addresses (street naming and numbering and development control) which are crucial to the effective creation and maintenance of the LLPGs. In this role as local custodians of LLPGs, the local authorities are the key to the creation and maintenance of the NLPG.

To facilitate the creation of the NLPG a British Standard (BS7666) has been created. The standard comprises four parts covering street gazetteers, land and property gazetteers, Addresses and Rights of Way. An operational BS7666 compliant LLPG is an essential prerequisite for involvement in the NLPG. BS7666 specifies a standard format for holding details on every property and street. The standard does not differentiate between commercial or residential properties, between occupied, developed or vacant land, between urban or rural or between addressable properties and non-addressable entities such as communications masts. The majority of local authorities in England, Scotland and Wales are linked to the NLPG, but despite the obvious benefits this standard has not yet been widely adopted in the commercial property sector although it is being used by some of the property information service providers eg Focus.

PISCES (Property Information Systems Common Exchange Standard) is an electronic data exchange standard specifically developed for the UK property market. This is based on the XML (Extensible Mark-up Language) technology standard designed for sharing virtually any type of information over the Internet. PISCES defines a standard format for transmitting property related data electronically between key business areas. It offers the property industry much greater flexibility in service procurement and the prospect of seamless transfer of data within and between systems and organisations. Software providers can have their software tested for compliance to the PISCES standard for data import and export and details of the standard and of compliant software is available at *www.pisces.co.uk*.

With the above in place effective benchmarking can be more easily introduced and managed. This is a valuable tool of investors, property managers and corporate occupiers as it enables them to monitor key performance indicators for a particular property against the standard for that type, and the bigger the dataset the more detailed the analysis and the more reliable the benchmark. A number of major property investors subscribe to the Investment Property Databank (IPD) which provides independent market indices and portfolio benchmarks for the property industry www.ipindex.co.uk. The Occupier Property Database (OPD) provides a similar function for corporate occupiers www.opd.co.uk. The PISCES format for data transfer is routinely used to transfer the data from the managing agents database to the IPD or OPD accurately and reliably.

Statute law

IT systems need to take account of current legislation and maintenance contracts should where possible provide for updates required by changes in the law. Legislation can be criminal, civil or, with a similar effect, requirements of a professional body like the RICS. For example, property management accounting systems need to provide the functionality required to conform with RICS client money rules.

It is useful if systems are sufficiently flexible or can be enhanced to incorporate the requirements of any relevant new legislation. For example, the Money Laundering Regulations 2003 requires client verification checks before specific types of instructions can be accepted. In order to prevent multiple requests for checks of the same client, this required systems to be enhanced to record the verification status of each client. Systems could be further enhanced to alert users to unverified clients and prevent billing of instructions until verification has been completed.

Conclusion

The role of the IT department has changed from process management through data platform to the shaper of the business environment. Accurate current, verified and accessible information now enables clients and advisors to devise new relationships for delivering services but systems are only successful when they are combined with effective business processes. This is resulting in fewer professional service providers, each with longer and more comprehensive contracts, often linked to outsourcing. This has implications for staff training and the provision of IT support and services across large networks.

The commercial property market has very specific requirements and the technologies and business practices continue to develop. A good up-to-date source can be found at the annual property computer exhibition *www.pcs-expo.co.uk* whose catalogue provides a directory of the majority of UK software suppliers for this market.

Management of the department

There are two aspects to consider. First, the environment in which the department can flourish within a larger business and, second, the operation of the department to provide property services.

Whether in-house or a consultancy, the normal business environmental requirements need to be provided:

- business management and development
- finance
- personnel and resources including information technology.

The business plan will differ between an in-house department and a consultancy. The in-house service is circumscribed by the property portfolio requirements of the specific public or private sector organisation. The main issues will be the means by which business planning interacts with property and the balance between providing the service in-house or through consultants.

As an example, in 1993, Legal & General defined the three levels of expertise required to run their portfolio as:

- asset management
- market orientated property work
- property management.

They spent £5 million a year on professional fees, mainly for market work, which includes:

- acquisitions
- marketing
- valuation
- rent reviews.

Whereas the consultancy is part of a larger practice, in King Sturge it is based on seven core values.

- *Integrity*
  To put the interests of our clients first and to act ethically, honestly and reliably.

- *Accuracy*
  To deliver precision in every aspect of our work.

- *Professionalism*
  To provide efficient and expert advice based upon extensive professional knowledge and experience.

- *Consistency*
  To ensure a coherent and harmonious service across our domestic and international networks.

- *Innovation*
  To anticipate the needs of our clients and provide solutions that are enterprising, timely and pragmatic.

- *Communication*
  To be accessible to clients and communicate effectively both internally and externally.

- *Fellowship*
  To enjoy what we do through a positive and supportive working environment.

These core values create the culture within which each of the departments develop and change in response to external factors.

The management department is concerned with:

- *practice management*
  - business planning and development
  - personnel
  - finance

- *managing clients*
  - obtaining new clients
  - building and developing relationships
  - agreeing contracts for professional services
  - reporting and accounting

- *managing property*
  - types of property

- *levels of management*
  - types of tenure and resulting technical issues.

These functions will be allocated in an hierarchy of partners, associates and surveyors. In large estates departments it is possible to distinguish:

- *Partners/directors*
  The management of the business in operational, financial and corporate terms with personal accountability and reward. This will include business development and the recruitment, development, training, reward and motivation of staff. Most important of all, responsibility for the top clients (whether internal or external).

- *Associates/senior surveyors*
  The management of portfolios and day to day liaison with clients.

- *Surveyors*
  The management of properties.

A quality service requires:

- a regularly updated management manual/QA system
- effective supervision and monitoring of surveyors
- demonstration of high quality work by senior staff
- close liaison with accounts staff
- mutual confidence between clients and staff.

In managing the practice, careful consideration will be given to the crossflow of work between the management department and market and consultancy departments.

A too-rapid turnover of managements can be damaging to the effectiveness and financial performance of the department. Clients of the management department arise principally from two sources, either as the natural result of development, investment or agency work, or from a competitive tender including the increasingly common practice of a review of present arrangements. Management's clients have inevitably a close and continuing relationship and, if a substantial number of properties are involved, then there is more to the task than a sum of those individual managements. The managing agent has the opportunity to contribute to achieving the portfolio requirements of the client whether they be expressed in investment terms or operational asset terms.

Further, each client is a source of income arising from both management and related work. Analysis of the cost of providing the service and direct and indirect fee income is essential in order to be able to quote for new work and respond to a formal tender.

The work from management clients include:

- occupational agency disposal
- investment acquisition and disposal
- building surveying
- asset valuation
- rent review
- planning and development
- rating of empty property
- performance measurement
- research.

The management of the portfolio of properties requires a wide range of skills:

- strategic thinking
- technical skills and knowledge of law and procedures
- attention to the detail of service charges and lease clauses.

In a large department this will be best achieved by separating out these functions. In one large practice, the role of asset manager (the portfolio manager) has been distinguished from that of the property manager (see below).

- *Asset manager*
  - Drafting and agreeing property management strategies and policies with investment partner and client representatives. Client liaison and business development with key assigned clients.
  - Maximise growth in property performance at rent review and lease renewal. Agree target rents and co-ordinate rent review negotiations. Prepare recommendations of any settlement to the client or the main client contact.

- Identify, recommend, exploit hidden values and development opportunities.
- Co-ordinate and prepare formal client reports and accounts.
- Management of team dealing with the effective day to day management of the properties. Overall responsibility for ensuring the timely service of all notices. Responsible for the financial management of service charge budgets and credit control procedures.
- Briefing of other departments involved in providing the property management service.

- *Property manager*
  - In conjunction with asset manager, maintain effective liaison with assigned clients and tenants.
  - Inspection of properties at regular and agreed intervals.
  - Supervise credit control of allocated client property portfolios in conjunction with the accounts section.
  - Maintenance of property and preparation of schedules of dilapidations and conditions (in conjunction with the building department).
  - In conjunction with asset manager, maintain effective liaison with assigned clients and tenants.
  - Assist with the preparation and co-ordination of the formal client report and accounts.
  - Supervision and training of site staff.
  - Prepare draft service charge budgets at the appropriate time on multi-occupied premises.
  - Implementation of service charge budgets.
  - Authorise invoices and expenses relating to the property. Deal with licences for alterations to property. Obtain tenders and deal with refurbishment of properties. Prepare draft valuations for fire insurance purposes.
  - Ensure obligations under health and safety and all relevant property legislation are being met.

Lastly and more generally, the structure of the firm and degree of delegation is a function of the work that is undertaken, for example:

- "brains projects" involve complex client problems requiring a high level of technical or professional skills

- "grey hair projects" involve carrying out familiar tasks involving some senior staff or partner endorsement but where some of the job can be delegated
- "procedure projects" comprise routine tasks that can be systemised or industrialised with the benefit of measurable quality control.

Thus, firms with different types of work will each adopt different structures. However, each will be involved in a process of achieving an agreed strategic plan with a structure designed to provide consistent client satisfaction which will require well trained and motivated staff with seamless departments, in operational terms.

All the above help to define the key features of the partners/directors/senior managers who have three distinctive roles:

- business entrepreneurs, creating opportunities and resources
- managers across a breadth of activity
- experts with a depth of knowledge.

While some are capable of performing two of these roles the author, in 40 years in the profession, has yet to find anyone capable of all three. Excellence in one is sufficient, which leads to questioning whether medium-sized businesses are ever likely to perform all three well.

Lastly, since 2003 many firms have started to consider the significance of Corporate Social Responsibility — how companies manage their business processes to produce an overall positive impact on society, or how the business recognises, communicates to shareholders.

9 Acquisition, Disposal and Lettings

Motivation of the parties

The buying or selling of property may be an isolated transaction in the life of an individual or of a small company but, in a larger corporate body, the frequency of such transactions will lead to these being conducted within formal procedures as part of the general management of the organisation. Before examining these procedures and their consequences, it is worth considering the motivation of purchasers and vendors. In very few circumstances, is it ever essential to sell or buy a particular property; the decision to initiate the transaction will usually develop as a result of a careful analysis of many factors and a number of possible properties may be identified, a transaction in any one of them being all that is required.

Since one man's acquisition is another man's disposal, the best transaction is one where both vender and purchaser are well satisfied. That is to say, having regard to all aspects of the property, the vendor considers he obtained a good price and the purchaser believes he has paid a reasonable sum. This can arise through various circumstances, partly as a result of different opinions as to the future of the market in general and the particular property. In the investment market, the existence of specialist management expertise, the potential of marriage value, an over concern with historic costs or the need to change the balance of a portfolio can all act as the catalyst to initiate the sale. The mere pressure of functional requirements accounts for most transactions by tenants and owner-occupiers adjusting to technical and economic change in their sphere of activity. Subject to the constraints of specialist users, the rate of change in relation to the financial structure of the occupier's business will determine the emphasis between renting and owner-occupation of either fully developed sites or new development.

Prior to the commencement of development, there may be years of site-assembly and holding management before the complex of negotiations, design and financing can be presented as a coherent and feasible package, meeting the requirements of investor, occupiers and public sector controls.

The client is represented by a team of advisors and their selection, monitoring, cohesion and supervision of areas of responsibility requires considerable personal skills on his part. They, for their part, will be involved in negotiations and discussions with public and private bodies and in their reporting may be influenced by factors other than solely the successful negotiation of the current transaction. The background, discipline and personality of the client's representative will influence the overall conduct of the negotiations and the significance attached to the opinions of the advisors in the various disciplines.

In addition, statute, in the form of the Estate Agents Act and Property Misdescriptions Act regulates and controls the practice and behaviour of property agents. If dealing with funding or insurance matters, the agent will also need to have regard to the Financial Services Act 1986 and subsequent amendments.

Acquisition

Acquisition can be for the purposes of development, occupation or investment, the same general process is involved for each of these:

- policy, leading to brief and then instructions
- search for suitable property, site finding
- preparation and investigation of short list
- detailed investigation
  - physical and environmental appraisal (possibly including performance analysis)
  - acquisition
  - finance
  - taxation
  - legal

  - management
- agreement, subject to contract
- recommendation and decision process
- approval and financing
- documentation
- implementation
- entry to portfolio
- physical and tenurial adjustment
- user
- management and monitoring.

Where properties are being acquired regularly then this process will itself be monitored and supervised to ensure that the best use is made of the financial resources and that analysis of the extent to which the brief has been met is fed back to the policymaking unit.

Where occupation is involved, particularly the occupation of a large unit, with considerable employment generation, then not only are political and social factors involved but the occupier will be interested in the general environment and infrastructure associated with a particular location. The weight be attached to these and the trade-off between individual properties and the locational features they provide are not capable of precise measurement.

Government programmes of providing financial incentives to companies in order to encourage development in depressed regions have proven to be less permanent than the features they were designed to overcome. Indeed, the Conservative Government of the 1980s all but abandoned the policy of regional support and left individual ventures such as the steel works at Ravenscraig to compete in the free market, struggling against the geographical factors the government had earlier sought to override. A subsequent attempt in the noughties to grant exemption from Stamp Duty Land Tax was similarly abandoned in 2005.

The range of issues which need to be considered when advising clients in a comprehensIve way on location and alternative forms of tenure are wider than those normally provided in property reports, but well within the expertise of the surveyor.

A potential occupier can buy or rent an existing unit, take a pre-let from a developer or develop to meet his own requirements. The latter is particularly true of the public sector and the more specialised types of occupiers in the private sector.

A potential investor can buy an existing investment, develop on a pre-let reflecting to some extent occupiers' requirements or develop speculatively.

Both occupiers and investors increase the scale of their activities and their property holdings by take-overs and mergers. Local authorities, institutions, property companies, retail chains and manufacturing companies are all involved and in the wake of such activities comes rationalisation and reorganisation of function and resources. There is a tendency to seek new and more centralised offices for the headquarters activities. At the same time, the operational assets will need the most careful analysis and review to ensure that the accident of ownership of the aggregate estate does not become the justification for continued use.

In the late 1980s, companies who were "rich" in terms of property assets but whose share value did not reflect this wealth were targeted by companies who sought to sell off their property assets. Those companies who did not frequently review their land holding found themselves vulnerable to "hostile" takeover bids. This pattern has been repeated as land values have risen and companies have failed to reflect this in their accounts.

Land-holding

Acquisition and disposal are, in many circumstances, quite unrelated; only a minority of acquisitions are made for the specific purpose of a subsequent disposal as part of a process of land assembly, with or without the provision of infrastructure. This is one of the functions of the builder/developer and is in its most sophisticated form in the case of residential land banks which seek to:

- ensure continuity of production
- allow a flexible response to changes in market demand
- provide for unforeseen delays on other sites
- achieve a partial monopoly in certain areas
- provide benefit from rising land values.

Many larger builders consider that three years supply is a normal requirement, but some land can be in the bank for many years before being developed. One of the major complaints of builders is that local authorities are not granting consents on enough land. This is partly due to the use of different criteria; the local authority tends to look at the aggregate land supply rather than the patterns of ownership of individual builders. A builder assesses each site in the light of its physical and financial suitability for his type of development, whereas the planning authority lack the incentive to consider the financial feasibility of sites which are available in a physical sense.

Research has shown that, in many cases, councils have overstated the land that is practically available by 15–20% in aggregate, which means by very much more in specific locations.

Where the land is being acquired for development, then the outline or detailed consent will normally be the single most important factor in determining the price of the land or, indeed, whether the land is purchased at all. A purchaser/developer will normally follow a mental planning checklist.

Purchaser's planning checklist

- Existing history of planning consents — were they implemented/still valid? Any conditions on existing consents?
- Existing use of the site. Certificates of Lawfulness of Existing Use and 1987 Use Classes Order (and subsequent amendments)
- Examine local plans and structure plans. What is planning policy background?
- Are there any listed buildings?
- Is the site within a conservation area, AONB, SSSI, etc? Are there any Tree Preservation Orders?
- Road proposals in the area.
- Do other planning consents in the area affect the subject property.
- Check enforcement register for outstanding enforcement notices.
- Lawyers to check for existing s.106 agreements and undertakings.
- Relevant circulars and Planning Policy Guidance Notes (PPGs) (now PPS).
- Contaminated land.
- Building energy certificates.

Planning obligation/agreements

Very often the negotiations for the acquisition of a site for development will take place in an environment in which the planning authority seek some planning gain. The reasonableness of the obligations can be tested by:

- Is the planning gain needed for the development to proceed? In the case of financial payments, will they assist this purpose?
- Is the gain clearly related to the development?
- Where related to a mixed development, is the condition imposed to secure an acceptable balance of uses?
- Is the condition related to the scale of the development?

A developer is entitled to refuse unreasonable conditions and appeal to the Secretary of State. Indeed, in *Westminster Renslade Ltd* v *Secretary of State for the Environment*, 1983, the High Court quashed a decision of the minister refusing consent because no planning gain was offered to the local authority. The inspector had refused consent because insufficient car park space under public control had been provided.

The 1991 Act introduced the concept of a planning obligation which encompassed both planning agreements and unilateral undertakings. These provisions amended s.106 which, in turn, replaced the old s 52 provisions of the 1971 Act. Basically, a planning obligation will constitute a legal device for restricting the development or use of land in a specified way. As with planning agreements, a planning obligation will be enforceable against the person who enters into the obligation and successors in title. The subsequent planning legislation has continued this policy.

Disposal

There is a natural inertia against the disposal of interests in property. There are many reasons for this, some explained perhaps more by psychology than estate management. In large functional organisations, including statutory undertakes, it may be difficult to find an individual prepared to say, "We will never need that property again", and if you take the alternative approach you can never be proved wrong. In the financial institutions disposals average around 2 or 3% pa of the total value of the portfolio. The measurement of portfolio performance is introducing a new discipline and thus more incentive to see disposal as a normal part of portfolio management.

Methods of disposal

The alternative means of marketing interests, private negotiations, auction or tender, each possess particular features which make them suitable for different circumstances.

Sale by tender

When the market for property is unstable or the vendor wishes to exercise some control over the identity of the purchaser, tenders can be invited, possibly in excess of a specified figure. If the site has development potential and the disposal is linked to a design brief then the potential purchaser's bid will be a combination of price and detailed scheme. Unless they have access to considerable development expertise vendors have difficulty in determining which of the combinations of price and project should be accepted. Even if the best proposal can be identified in a quantitative way, the purchaser's ability effectively to implement the scheme and manage the overall project needs to be carefully considered in the light of their track-record. The alternative approaches include a preliminary tender and then negotiations with several short listed potential purchasers, which gives an opportunity to assess their performance. Tenders can be abused by purchasers who use them to get "their foot in the door" and then steadily renegotiate the terms.

However, a tender provides the opportunity for one fixed specific bid only by a certain date, made in ignorance of any other bid by any other party; such a bid can only be the highest the purchaser is prepared to offer. The vendor is not, of course, bound to accept the highest tender and there are circumstances when it would be prudent to accept a lower bid, and this possibility should be made clear in the tender documents.

Sale by auction

While tenders can result in a higher price than that obtained by other means, some potential purchasers may indicate unwillingness to take part. An auction is organised to realise a price one bid over that of the second highest bidder and in private negotiations there is sufficient feedback of information and opportunity to make a series of bids for the purchaser to ensure that he pays no more than he need to obtain the property.

A well organised auction should absolve any vendors acting in a trustee capacity from criticism as to either their general conduct or the price realised. The date, venue and form of the auction need to be carefully planned to prove attractive to what are seen as the typical purchasers, avoiding bank holidays and other holiday periods. The benefit to a vendor is that a successful bid is immediately followed by the signing of a contract and the payment of the deposit, but this could deter some potential purchasers when credit is difficult to arrange.

In addition, a vendor may not favour a sale by auction in a situation where he wishes to be able to exercise control over who purchases a property. This could be awkward in an auction room where a bidder has clearly outbid his rival but where a vendor, for whatever reasons, does not want to sell to that particular bidder.

The recommendation of a reserve by the auctioneers is never an easy task, and even then the client may issue other instructions. A sale at a price well in excess of the reserve reflects poorly on the auctioneer's knowledge of the market; failure to reach the reserve is possibly indicative of other problems.

In 1992 Jones Lang Wootton launched a quarterly auctions results analysis system. The first annual update in January 1993 showed a shrinking market in terms of both the number of lots offered and money raised. The success rate in 1992 (out of 1,156 lots offered) was 68–70%. In 1991 the same auctioneers offered 1,711 lots and had a success rate of 73%. These results, coming at the bottom of a severe recession, compare poorly with those of the 10 major firms in 1984, where the success rate varied from 75 to 91%, with most of the firms reporting a success rate of about 80%. Since then, results have fluctuated according to the particular economic climate and market sentiment at the time but have generally increased towards the 1980s levels.

In a depressed market, few vendors will want to sell by auction as an auction requires visible competition in the auction room — competition that might not be forthcoming. In a stronger market, the auction method of sale will be more popular as the competitive atmosphere encouraged by the auction room may encourage bidders to exceed their own pre-auction bid levels.

Sale by private treaty

The vast majority of properties are sold through private negotiations but the period of time involved can vary from hours to years. In the case of prime investments, the relatively small number of surveyors that act on behalf of the institutions and property companies rely on personal knowledge of those in the market. What would appear to be a permanent shortage of the prime property investment means that for much of the time a seller's market exists as institutions seek to place part of their continually growing income into these properties.

The advantage of the private treaty sale method is largely one of convenience and flexibility, there are few controls or restraints over procedure (apart from the Property Misdescriptions Act) and it is generally favoured by purchasers, particularly in a depressed market.

The most commonly quoted disadvantage is the apparent lack of any fixed "closure" date, leading often to protracted negotiations and revised offers. In practice, many agents seek to avoid this by inviting informal "best bids" by a particular date, giving a sense of urgency to the sale, but without adopting the formal tender procedures.

Best method of sale

The selection of the best method for a particular sale and its effective implementation is a function of the character of the property and the motivation of the vendor. In the case of a public sector client, there may be a preference for auction or tender and the agent involved should expect to keep a more formal record of all aspects that are involved in working towards a disposal. No method is exclusive; an unsuccessful auction is often followed by a private sale, and a sale where offers are invited in excess of a specified sum can sometimes only be resolved fairly between several potential purchasers by the introduction of an informal tender procedure.

All disposals require a carefully planned advertising programme using the various alternative media to publicise the property to best advantage.

The agent will need to consider from what group the eventual purchaser will come and how best that group can be targeted. The method of sale recommended will be of little importance if the prospective purchasers are unaware of the availability of the property.

Common to most marketing campaigns will be the following elements:

- preparation of sales particulars
- advertising campaign in media
- advertising on the vendor's agent's website
- mailshot to registered applicants
- mailshot to other possible purchasers- local companies, etc
- press releases to media
- follow up of initial enquiries.

Sales where equity is retained

Not all disposals involve a straightforward transfer of title. A landowner may wish to retain an interest in the property until some future date or event. The most common schemes include the following.

Joint ventures

The land owner and developer form a "partnership" to develop the site jointly. Particularly favoured where the land owner wants either to share the risk of developing land with another party, or where the land owner does not have the resources to develop the land.

Option agreements

Developers agree an "option" to purchase land at a future date, often triggered by an event such as the grant of planning consent. A formula or fixed price is agreed to determine the eventual level of the transaction, with often an "option fee" paid in advance by the developer. Ideally suits residential developers who require substantial land banks, but who wish to avoid substantial up front outlays.

Conditional contracts

Many purchasers will not agree to a contract until uncertainties (such as whether planning consent will be granted) have been eliminated. By agreeing to a contract conditional upon certain events, the vendor and purchaser are able to agree the principle terms of the transaction and reduce uncertainty by contractually committing themselves to the transaction so long as certain events occur.

Right of pre-emption

The grant by a land owner of "first choice" to another party (perhaps a tenant) in the event that the property should, at a future date, be marketed. Such a right is often negotiated as part of a lesser interest in the property.

Development sales and acquisitions

In recent years development has been accepted as not only an activity but a study in its own right, and this is reflected by a growing volume of research, analysis and information, the existence of which has permitted this text to leave unmentioned the effect of the various factors which contribute to the development process. There are, however, some particular types of transactions within the development process which are an integral part of estate management.

A ground lease, typically for 125 years, represents an ideal way for development to occur within a comprehensive framework controlled by the ground landlord. There is a natural tendency for the developer to seek a better package of terms from his occupying tenant than he is prepared to agree with the ground landlord. To some extent this is essential, since without the resulting profit there would be no developer and no development. The return which the landlord receives is a ground rent, reviewed at certain intervals, related to the rents achieved for the completed building, typically between 10 and 20% of the full rent.

Rather than agree a fixed review period for the ground rent of say five years, a landlord can seek an "as and when" basis, that is to say, as occupation rents are actually reviewed or renewed, so the ground rent rises in proportion. Clearly, this will result in more complex accounts but it is likely to lead to a more attractive cash flow and as occupation leases terminate and shorter frequency rent reviews are agreed, this is automatically reflected in the frequency of ground rent reviews. To what extent it is necessary or possible to agree an overriding provision to protect the landlord against any downside risk is a matter for negotiation.

In *Freehold and Leasehold Shop Properties Ltd* v *Friends Provident Life Office*, 1984, the rent payable to the ground landlord was 17.5% of; the rack rent (if the property was let as a whole) or the aggregate total of rentals payable in respect of lettings current at the review date and agreed rents in respect of any parts then vacant, whichever is the greater. It would be in the interests of the head lessee to negotiate with his underlessees payment of a premium in lieu of rent, hence reducing the payment to the ground landlord to a minimum.

Public sector disposals

The normal market factors which influence disposals are complicated by several other factors in the public sector. In both central and local government there are political issues and the Conservative Government elected in 1979 had as strong a political will to dispose of land as the Labour Government elected in 1974 had to acquire land. There are also technical aspects — the scale of operation is large, having economic effects beyond those of just land transactions, and the accountability required of the public sector may require some particular procedures to be followed.

The political philosophy behind the Conservative Government's programme of disposal has been explained by the Secretary of State as based upon a number of objects that coincide as follows:

> first we have a very large public sector borrowing requirements and to the extent that we can release assets we are able to reduce the borrowing requirement. Second, we are trying to ensure that assets are used as effectively as we can, particularly in terms of land. Therefore where the public sector has got land that is not used as effectively as we would like, we are very anxious to ensure that it is put on the market and sold to people who have a use for it. Third, we are determined to change the balance between the public and private sector by extending the private sector. So for all those three reasons we are examining the assets in State ownership with a view to disposing of them where we think it practicable and desirable to do so.

The subsequent Labour Governments have generally continued this policy.

Whether that philosophy appeals or not, its implementation by surveyors in the public service and consultants represents a strategic operation and presents new problems.

The scale of the disposals raises a number of issues which last arose on the sale of land surplus to requirements after 1945. The *Critchel Down* case concerned the sale of agricultural land acquired for a bombing range before the war which the previous owner discovered had been sold to the Commissioners of Crown Lands. After a lengthy campaign to repurchase the land the minister made a statement in Parliament in 1954 that:

> The Government has now made the important new decision that where a former owner or his successors can establish a special personal claim, he will, where possible, be given a special opportunity to buy back the land

> when it is no longer wanted for Government purposes ... at a price to be assessed by the district valuer as being the current market price.

The Critchel Down rules are not statutory but are commended for action by public bodies. They also apply to the private sector to which public sector land holdings have been transferred, for example they apply to the water authorities privatised in the late 1980s. In situations where the rules apply the former owners are given the opportunity to buy back their previously owned land.

There are some limitations on the scheme; if the land has been materially improved since acquisition the rules do not apply, where, for example, houses have been erected on agricultural land; mainly open land has been afforested or offices built on an urban site.

The Critchel Down rules have evolved since their inception in 1954. For example, in 1992 the Under-Secretary of State modified the rules as follows:

(a) qualified valuers other than the district valuer should be able to carry out valuations for disposal
(b) there should be an additional exemption from the offer back obligation in cases where competitive sale is essential to establish market value
(c) the periods for allowing former owners to indicate an intention to purchase and agree terms should be simplified by allowing two months in all cases for indicating an intention to purchase, two months to agree terms other than value and a further six weeks to agree the price;
(d) in the case of future acquisitions, a 25-year cut-off on the obligation to offer back to former owners should apply to agricultural land as it already does to other land
(e) the terms of sale should include clawback provisions in cases where the planning position of a site is unclear.

The provisions for clawback outlined in (e) above reflect a serious concern that public sector land should not be seen to be undersold. Although not covered by the Critchel Down rules, the sale of the Royal Ordnance by the government in 1987 led to such concern that the Treasury revised its guidance notes on the disposal of public land and buildings. British Aerospace and Trafalgar House purchased two north London former Royal Ordnance sites from the government — Royal Ordnance sold the sites for £3–5 million, a valuation made on the assumption that their location in the green belt gave them less development potential than was actually the case. Apparently, the valuation was based on a surveyor's estimate in 1984. In 1989 the value of the site as a mixture of industrial, commercial housing and leisure elements was estimated at around £400m.

In July 1992 HM Treasury revised its guidance to public bodies to bring it up to date with lessons learnt from the Royal Ordnance and other sales. Among the revised provisions, the following were particularly relevant to Royal Ordnance:

(a) ... to ensure that in seeking professional advice on the sale of property they always ask for advice on whether there is likely to be potential for development
(b) ... to consider a "clawback" provision in cases where land is disposed of before uncertainties about the planning position have been satisfactorily resolved
(c) ... to seek advice on the circumstances when a valuation independent of the selling agent should be sought.

Another consequence of the Royal Ordnance sale is that very few sales of public land are by way of private treaty. Sale by tender or auction often remains the preferred route, also providing a clear public demonstration that the land has been sold openly at the best possible price.

In June 2004, the ODPM published a circular, which superceded the 1992 Crichel Down Rules by limiting the requirement to offer back property to the original owner

The disposal of land by local authorities is referred to in s 123 of the Local Government Act 1972. A general disposal consent order of 2003 provides that Secretary of State approval is now only required when an authority makes a sale as an undervalue greater than £2m.

Marriage value

The concept of the whole being worth more (or less) than the sum of the parts is not just a feature of property. In industry, mergers between companies are often justified on the basis that the two companies involved each possess specialist technical, financial or marketing skills of a semi-monopolistic character which would, if exercised jointly, be to the mutual benefit of both companies. In the event this may not always prove to be the case. Similarly, in personal terms, a group of individuals with a common purpose may form a coherent team or a bickering committee. In real property, marriage value consists of tenurial and physical aspects and both can be present to varying degrees. Consider the successful and well-established tenant farmer who has the opportunity to buy the freehold vacant possession interest in land adjoining his tenanted farm. Because he can farm this with no increase in his capital equipment, his fixed cost can be spread over a larger acreage and that land is worth more to him than to some other potential purchasers because of his existing leasehold interest. In fact, the new land need not adjoin his present farm but merely be conveniently located. At the other extreme the developer with 99% of a carefully assembled site will be very anxious to acquire the final outstanding property and only he will know its true value, having acquired, ideally, a freehold interest. The scheme usually requires the demolition of any buildings on the site.

The concept of marriage value has been recognised in statute and case law for many years in the fields of compensation and taxation, for example, in the formula used to adjust costs on the part disposal of an interest for capital taxation purposes.

Tenurial marriage value generally arises from the differing approaches for the valuation of freehold and leasehold interests. As long as dual rate tax adjusted tables are used to value leasehold interests and single rate tables for freeholds, then the combination and merger of the component interests will result in a higher value than the aggregate of the component interests. The distribution of the surplus between the parties will depend upon their relative negotiating strengths and perception of the position of the other parties. The extent of this is most marked in the residential field, where tenants enjoy security of tenure and either party on obtaining the right to dispose of the property with vacant possession can make a substantial and possibly tax-free gain. In the commercial field the tenant in occupation may also be prepared to offer more for the freehold or long leasehold interest in his property than would be obtained if it were sold as just one of a number of properties in an investment portfolio.

The break-up of an interest can result in greater aggregate proceeds than if the whole interest were sold as one lot, though consideration must be given to the extra time and costs involved. An investment consisting of a very large, fully developed site comprising a number of distinctive buildings might be worth more if available in the market as several lots. The principle was articulated in *Ellesmere* v *Inland Revenue Commissioners*, 1918, in connection with the break-up value of an agricultural estate. This gave rise to the phrase "prudent lotting" as one of the valuation rules for the then estate duty and is equally relevant to inheritance tax valuation. To the extent that this is the opposite of marriage value, it can be called divorce value.

The rearrangement of interests and the terms upon which they are held is the practical application of marriage value concepts to detailed portfolio management. An important feature of a commercial investment is a "clear" lease, all the outgoings being the direct responsibility of the tenant or at least the financial responsibility through a comprehensive service charge. It will be in the landlord's interest to offer his tenant who holds on old type lease terms, some encouragement to alter the terms at review or renewal.

Letting

Carefully selecting and screening prospective tenants is essential to ensure that the rent is likely to be paid on time and to avoid the costs of tenant default by way of:

- unpaid rent
- voids
- breach of covenants
- reletting costs.

While government or corporate tenants come with guarantees (assuming that it is the parent company taking the lease), smaller private sector tenants need to undergo a systematic process to establish their general suitability and covenant strength. This would testing and checking their information. A vacant property may be worth more than a bad tenant.

Incentives

When markets are poor for landlords, it becomes necessary to offer incentives to potential tenants. There are many different letting incentives which differ in their attractiveness between landlords and tenants including their liability to tax. The most common are:

- rent free periods, creating the difference between headline and effective rents
- fitting-out allowances

- capped service charges for a specific period
- options to break or renew for a variety of reasons
- capital contribution by landlord to boost rental, ie "reverse" premiums
- undertaking of improvements with capital or rental adjustments
- "affordable rents" are sometimes offered to take into account a specific type of tenant's inability to pay a full open market rent eg in order to improve the tenant mix of a shopping centre.

Advisors need to ensure that the value to their client exceeds the cost, that the agreements are fully documented and the consequences as regards capital and income taxes and VAT have been fully considered.

Energy performance legislation

Arising from the 1997 Kyoto Protocol, a directive on the energy performance of buildings became part of European Community Law in January 2003, becoming part of UK law by January 2006, although some of the requirements may be deferred. It affects almost all building owners and occupiers, who will be required to have an energy performance certificate available for inspection by anyone wishing to purchase or lease the property.

The directive includes four key areas.

1. *An energy performance methodology*
   A methodology must be applied to calculate energy performance which is used to set minimum energy performance requirements to reflect issues such as building function, age and general climate conditions. Differentiation can be shown between new and existing buildings but most types of buildings are included.

2. *Building energy certificate*
   Upon any sale or letting, an energy performance certificate, no more than 10 years old, must be shown to the prospective purchaser or tenant. A certificate is must also be produced for the owner of a newly constructed building. A single certificate may be produced for a block containing apartments or units for separate use, where the block has a common heating system or, alternatively, there can be an assessment of a single representative apartment or unit.

The certificate must also include recommendations for cost-effective energy improvements and may, for larger public buildings, provide for the display of current and recommended indoor temperature.

3. *Minimum standards in new and existing buildings*
   New buildings will be required to meet minimum energy performance standards. Where the design of a new building exceeds a total useful floor area of 1000m^2, the technical, environmental and economic feasibility of on-site electricity generation from renewable sources must be considered.
   Existing buildings with a total useful floor area of over 1000m^2 undergoing major renovation must be upgraded to meet minimum energy performance requirements where it is technically and economically feasible. It is up to Member States to decide whether these requirements should cover the whole building or just those parts being renovated. Differentiation may be made between performance requirements for new and existing buildings.

4. *Inspection and assessment of heating and cooling installations*
   Inspection requirements for boilers burning fossil fuels, and air-conditioning systems are laid down with a view to reducing energy consumption and limiting carbon dioxide emissions. Inspections must be carried out by independent qualified and/or accredited experts.

Little practical action is possible until the full UK legislation has been published and procedures clarified, however it will have major implications for property owners and developers.

Tax planning

The complexity and rate of change of tax avoidance measures mitigates against a detailed commentary. The relatively recent introduction of Stamp Duty Land Tax, particularly the treatment of leases and the probability of REITS are but two examples.

However, there are some principles for the estate manager to consider when acting as client, obtaining advice and being a member of a project team.

Tax planning can be defined as:

> The holding, dealing, transferring selling, developing, letting, licensing or granting of options in property in such a way as to maximise the present value of the net proceeds after tax from the point of view of investor or occupier who may have the status of: individual, partnership, company or charity.

However complex the transactions may seem and baffling the work of the parliamentary draftsman most tax planning revolves around key concepts and assets, which set a framework for tax planning.

Key concepts

- The status of the parties, influenced by both their title and their conduct. For individuals the introduction of SIPPs and the 2006 changes are crucial.
- The character of the transaction, in the case of income, directing it towards the most favourable schedule and case to offset expenses.
- Exemptions and reliefs, particularly the case with capital taxes.
- Timing, which includes both the order of events and the period of time for which assets need to be held to benefit from exemptions and reliefs.
- Elections and notifications whereby the tax payer can choose a regime or nominate an asset.
- Interaction of taxes, which can arise with Capital Gains Tax/ Inheritance Tax and Stamp Duty Land Tax/VAT.
- Valuation, where regard must be had to "prudent lotting" and "flooded market".

Assets

Different assets have, over long periods of time, acquired quite distinctive tax treatment:

- agricultural land
- forestry
- business assets
- plant and machinery
- owner occupiers main residence
- National Heritage assets.

The availability of grants, particularly the comprehensive restructuring of grants from agricultural production to the quality of land management, between 2005 and 2015 has similar implications to taxation.

Occupiers

Business rates apply solely to occupiers. Quite apart from quinquennial revaluations and the right to appeal against the new assessment, for example the 2005 revaluation based on April 2003 levels of value, occupiers can pursue numerous initiatives:

- splits of the assessment can isolate a vacant or separately occupied hereditament
- mergers can bring together separate parts with a lower overall basis of value
- empty property, if it can be separately assessed, will be exempt from rates if industrial or a listed building
- temporarily vacant property may qualify for relief from rates
- in the contractors method of valuation some public assets have a lower decapitalisation rate
- a change in the description of property in the valuation list may result in a lower valuation or exemption
- a vacant commercial property for which it can be shown that the next use is residential can obtain relief
- a commercial property awaiting refurbishment may be capable of a nominal assessment, reflecting the very extensive nature of the work
- a material change in circumstances, infrastructure works, new buildings, disasters etc, can create grounds for a reduction in the assessment.

All the above require identification and effective and timely communication of the issue to a specialist surveyor who can assess the options and serve valid timely notices on the Valuation Office and negotiate a favourable outcome, based on comparables, experience, cases and statute.

Stamp Duty Land Tax

Since 1 December 2003 transactions relating to UK land have been liable to SDLT and are no longer liable to Stamp Duty. The tax is

Rates of Stamp Duty Land Tax

Residential property	*Rate*	*Non-residential property*	*Rate*
Below £150,000	Nil	Below £150,000	Nil
£60,001–£250,000	1%	£150,001–£250,000	1%
£250,001–£500,000	3%	£250,001–£500,000	3%
Over £500,001	4%	Over £500,001	4%

payable at rates which depend on the amount of consideration paid. If the payment is contingent — for instance on further planning permission being obtained — it will be assumed that the amount will become payable. If the payment is uncertain, a reasonable assumption as to the amount will be made. Where transactions occur between connected persons, the purchase price will be assumed to be not less than market value.

SDLT is paid by the purchaser, as at Spring 2005, the rates of duty were as shown in the table below. The rate is applicable to the whole of the consideration — progressive application does not apply. "Land" includes transactions that "relate to land", eg transfers of shares in land-rich companies.If conveyances are executed in the UK, duty will be payable, even if they relate to land outside the UK.

The purchasers of property have 30 days from the date of the conveyance to notify the Stamp Office, otherwise interest charges arise. Notification is required even where no tax is payable. The Stamp Office will provide certification, and this is required to register the transaction with the Chief Land Registrar.

Leases

Not only freehold transaction, but also certain lease transactions are liable to SDLT, notably lease premiums and rentals. Premiums are charged at the same rates as freehold transfers, unless the annual rent exceeds £600, in which case the nil band is replaced by a 1% rate. Certain lease variations are caught, including extensions to terms, capital payments by landlords in respect of reduced term lengths, and capital payments by tenants in respect of rent commutation. SDLT is levied on payments to tenants in respect of surrenders.

Rents are the subject of complicated new provisions. SDLT is levied on rents due under leases. A flat rate of 1% is charged on the net present value of the rental income for the duration of the lease, with rents being discounted at the RPI. The charge only applies to the excess of NPV over the relevant nil rate threshold.

For the purposes of the NPV calculation, any rent increases after the initial five years of the lease are ignored: subsequent rents are deemed to be at the highest amount paid during the initial five years. Turnover rents or other uncertain rents are based on reasonable estimates.

Caveats

Of all taxes VAT has had the most profound effect on property and it is essential that this aspect is clarified in any transaction.

The growth of large mixed-use development schemes adds another complexity in as much that different uses may be subject to different regimes which affects the development process, this is particularly so in respect of maximising the value of capital allowances.

Tax planning requires a relatively sophisticated client and a critical mass of asset to justify the time, cost and complexity involved and avoidance and evasion need to be carefully distinguished.

Lastly, the increase in tax at the point of a transaction renders property transactions less attractive, thus investors will want to see a larger benefit from the transaction or seek to transact the asset that holds the property, so pricing will tend to move away from property comparables to securitisation comparables.

Sources and further reading

Compulsory Purchase and The Crichel Down Rules, ODPM Circular 06/2004.

Critchel Down Rules Topics, Crowther, EG, 12 December 1992 (note that Crichel Down Rules amended in 2004).

Disposal of Land and Buildings, HM Treasury, 23 July 1992

DOE Circular 9/80, Land for Private Housebuilding.

Estates Gazette Auction Results pages.

VAT on Property & Construction, RICS, 2005.

10 Portfolio Performance

Introduction

This chapter is essentially about real estate portfolios, collections of interests in land and buildings, and issues relating to their performance. It begins with a brief introduction to the nature and classification of real estate portfolios, and an insight into what is meant by the term performance. This is followed by separate sections focusing on performance measurement and analysis of different types of real estate portfolios, namely investment and corporate (or operational) portfolios. While the focus is largely on matters relating to real estate, this is not possible without reference to a range of issues relating to the wider investment marketplace, consequently such issues have been incorporated in the discussion as relevant and appropriate.

Real estate portfolios

Interests in real estate, be they freehold or leasehold interests, or other rights over land, are held by a diversity of parties for a variety of reasons.

Parties that hold legal interests in real estate include the Crown Estate, the church, statutory bodies, financial institutions, corporate

---

Note: This chapter has been revised by Steven Tyler but remains based on chapter 11 of the third edition of *Estate Management Practice*. The inclusion in this chapter of material previously written by Professor Steven Hargitay and Wayne Miles, and which was published in the previous edition, is acknowledged.

organisations, private companies and individuals; while not a definitive list, the list certainly illustrates the eclectic mix of parties that hold legal interests in real estate.

Reasons for holding interests in real estate can be robustly classified into two main categories; occupation and investment. Whilst this might seem overly simplistic, for example some parties might simply hold interests in real estate as a consequence of historical legacy or for pleasure (which might not seem much like investment), or a local authority might hold and sublet land in order to ensure the provision of a community service — either by themselves or by a third party (again an activity that may not appear overly investment orientated in nature), the classification provides a working framework that accounts for a substantive proportion of real estate interests.

In situations where organisations and parties hold real estate primarily for investment reasons, there is often separation of the freehold ownership of the interest, and physical occupation of the land and property. Much commercial real estate (which for the purposes of this discussion is assumed to include retail, business, industrial, and commercial leisure premises), together with farmland and residential properties, are held by financial institutions and other investors, and leased in order to secure regular returns on capital invested, and perhaps also future gains in capital value.

Bodies and individuals that choose to occupy real estate generally do so in order to enjoy the rewards, benefits and opportunities of occupation. These may range from having a home, to providing premises from which to operate a business or service, for example manufacturing, retailing, financial or public services. While some occupiers may perceive and appreciate the investment potential of the real estate held, usually the primary motive for occupation of the real estate lies elsewhere, namely the desire to operate the particular business or service. Much real estate in the commercial sector is occupied by corporate organisations for which real estate is a factor of production and an essential business resource; this real estate is generally referred to as an organisation's corporate or operational real estate portfolio. In the ensuing discussion it is assumed that the terms operational and corporate real estate can be used interchangeably and no distinction is drawn between the two terms.

Whether the primary motive for holding real estate is investment or occupation, a large proportion of the commercial property sector is distributed across investors and occupiers who often hold multiple interests. Multiple holdings of such interests can be conveniently termed "portfolios"; either investment or corporate (or operational) real estate portfolios, depending upon the underlying purposes and motives of the holders of these interests. It is these portfolios, investment property portfolios, and corporate property portfolios, that this chapter is about.

Real estate performance

Performance can be defined as achievement relative to objectives or targets. If achievement can be measured, then the extent to which it has been possible to meet the objectives or targets set can be expressed in quantitative terms.

The natures of the objectives set for real estate investment and corporate portfolios, whilst sharing some common aspects, usually differ substantially, and present quite different practical and conceptual measurement problems and challenges.

For example, holders of both investment and corporate portfolios are likely to want to know about the location, structure, size and value of their portfolios. Holders of investment portfolios will particularly want to know about the financial performance of individual properties and sections of the portfolio in terms of rental and capital growth and return on capital employed. Holders of corporate portfolios are likely to be particularly interested in matters relating to operating costs, for example, rental payments, rates, heating, lighting, and maintenance. Portfolio managers of corporate real estate may be particularly interested to see such costs expressed in relation to units of resource, for example cleaning costs per unit area or per employee. Additional measures that may be of interest to the corporate occupier and operational portfolio manager are the "softer" measures of customer and employee satisfaction with the workspace.

However, as well as being interested in the nominal performance of individual real estate interests within the portfolio, holders of both types of portfolios will be interested in the relative performance of individual and groupings of interests, both within and outwith the particular portfolio. For example, how an individual property's performance compares with that of the portfolio sector or grouping based on particular attributes, and how the portfolio's performance compares with the performance of competitors' portfolios or other external benchmarks including indices, league tables, and performance statistics.

Clearly it is evident that performance measurement is not an end in itself, rather it is a means to an end, performance measures provide a platform from which to investigate and seek to better understand performance; as expressed by Hargitay (1986) "... measurement without subsequent analysis is worthless." Analysis of the results of portfolio performance measurement is aimed at developing a better knowledge and understanding of the underlying nature and reasons for performance. Equipped with knowledge and understanding of portfolio performance, the decision making capabilities and potential of investment and corporate real estate managers should be enhanced, leading to the prospect of improvement of the performance of the portfolio and enhanced operational effectiveness and efficiency.

Investment portfolios

In this section the following range of issues relating to real estate investment portfolios are examined: the increase in investment activity; investment needs of investors; portfolio management; performance measurement and analysis; particular difficulties arising; and market indices.

Real estate investment

Since the 1950s an unprecedented growth in investment in real estate has been observed; institutional and private investors have substantially increased the proportion of direct property investments in their investment portfolios. Since the mid 1980s, major changes have taken place in the Stock Exchange and new investment vehicles have emerged through the unitisation and securitisation of property; private investors have increased their commitment to real estate investment in a variety of ways. For example, investors have invested directly by purchasing homes for personal use, and second homes for leisure and investment purposes; this process encouraged by recent buy-to-let initiatives. Additionally investors have invested indirectly in real estate through pension fund investments.

The growth in this form of investment activity is based on the belief that real estate as an investment medium provides relatively good security of capital and income in an uncertain economic environment. Besides promising relatively attractive returns, real estate also offers excellent opportunities for diversifying large corporate and institutional investment portfolios.

The investment characteristics of real estate are significantly different from the characteristics of assets in other investment media. This is the reason why real estate is so useful and attractive for the purposes of diversification. On the other hand, historically, such differing characteristics isolated property from the other media, in which substantial advances were made in the development of decision-making methodology and the modernising of investment and portfolio management techniques. The full acceptance and integration of property into the global investment portfolio depends on the full understanding of the investment characteristics of property, not only in isolation, but also in the portfolio context.

The construction and management of real estate portfolios must be executed in an efficient and expert manner. The level of expertise of real estate investment portfolio managers must be on a par with that of their counterparts who manage portfolios of investment assets from other investment media. As the unitisation and securitisation of property assets continues to develop, and investment managers and real estate professionals extend their knowledge, understanding, and expertise, real estate will become further accepted and integrated into the global investment portfolio. Prerequisite to such integration is the recognition of the special investment characteristics and features of real estate, and understanding of the process of real estate portfolio construction and management. It requires the continued examination and evaluation of the theories, methodologies, and analytical decision making techniques which have been developed in other investment media, in order to assess whether they can be usefully adopted and applied, with or without adaptation, in real estate investment and portfolio analysis.

Considerable comment on real estate investments is in terms of prime property, demonstrated by four principal elements:

- the best location for the particular land use
- the best physical structure on the site
- the lease form and income flow should represent best current commercial practice
- the tenant offers the best covenant.

However, the prime characteristics are not intrinsic to a property investment as:

- the 100% pitch in a retail location can change as new development occurs
- a building can suffer wide-ranging obsolescence
- lease clauses develop over time and the best characteristics of one generation are the embarrassment of the next — as evidenced by the reduction in the length of leases, inclusion of break clauses, and the pressures to move away from the upward only rent review clause
- the stock market's opinion of trading companies can change quickly for many reasons, altering sentiment towards the quality of the covenant strength.

Investment needs and decision making

In a general sense, investment means the commitment or foregoing of resources, through the acquisition of assets by the investor, with the expectation of the return of a greater level of resources in the future. In a real estate investment context such future resources will usually be in the form of income receipts (rent), increases in capital value, or a combination of the two.

Investment decisions have traditionally been made in the light of the results of fundamental analysis which takes into account the state of the economy, the performance of a particular market sector, the financial and market status of the company, the valuation of a company's share, and the prediction of its expected future profits and dividends. Technical analysis, which involves the charting and analysis of historical price movements to predict future price movements, is often used to underpin the conclusions of fundamental analysis. Investment decisions are usually made on the basis of beliefs about the future, such beliefs often being based on past experience. Most investors accept the fact that their beliefs about the future are imperfect and all their investment activities will, inevitably, involve some risks. Rational investors will aim to achieve the best possible return for the level of risk incurred.

Investment decisions therefore, are not just based on expected return, as the capital committed to the acquisition of assets and the expected returns, are exposed to risk; risk being defined as the chance of adversity or loss, and the likelihood of not achieving the expected return. Generally, the greater the exposure to risk the higher the rate of return expected by the investor as a reward for bearing the risks involved.

There are a variety of different types of assets from which the investor may choose in an attempt to achieve the desired risk-return balance.

The main asset types available to the investor are cash, financial, and real assets; real estate being a real asset. The selection criterion of individual assets from these asset types is based on the provision of an adequate rate of return at an acceptable level of risk exposure; although it is argued that the rational investor seeking maximum efficiency in terms of the risk-return relationship will seek to maximise the level of return achieved for a given level of risk, or, minimise the level of risk exposure necessary in order to achieve a given level of return. Most rational investors are risk averse, in other words they prefer less risk to more risk, and more return to less return. However, whether in reality all investors act rationally is a point open for debate.

While in theory at least, most rational investors aim to maintain investment risks within certain parameters, the attitudes of investors to risk bearing is entirely subjective and very difficult to express in quantitative terms. A fundamental problem with risk is that it is difficult to define and even more difficult to measure.

The concept of diversification across a range of investment assets as a rationale for the management of investment risk is long established, the essence of this concept being recorded succinctly in the Babylonian Talmud as follows:

> Man should always divide his wealth into three parts: one third in land, one third in commerce and one third retained in his own hands.

In more recent times this ancient concept has become embodied in the portfolio approach to investment portfolio construction and analysis, a far more theoretical and analytical framework and model referred to as Portfolio Theory, developed by Markowitz (1952, 1959). In this context, the portfolio is a combination of several investments assembled for the purpose of the management of investment risk and for the enhancement of investment returns.

The seminal work of Markowitz was followed by the development of further theories and models by others active in this field of inquiry, for example the Capital Asset Pricing Model (Sharpe, 1964) and the Efficient Markets Hypothesis (Fama, 1970). These theories and this general field of study are now collectively referred to as Modern Portfolio Theory.

Modern Portfolio Theory provides the theoretical basis for the rational selection and assembly of a portfolio of risky securities or investments. It provides the rationale to select a combination of both risky and risk free assets that would meet the investor's goals and objectives. The first and fundamental problem facing the investor is to establish investment goals and objectives. Only after the clear definition of investment goals and objectives can the appropriate investment strategies and tactics be worked out. The formulation of the portfolio objectives involves the recognition and consideration of the principal needs and constraints of the individual investor. The generic needs of investors include security of capital and income, marketability and liquidity and taxation.

Security of capital

In most cases, security of capital, in both nominal and real terms, is likely to be one of the most important considerations, as the preservation of the value of individual assets and the portfolio is likely to be a prerequisite to investment performance; if capital is lost, its replacement can be extremely difficult, and the effect on financial performance disastrous. An unplanned total loss of the capital is usually the result of recklessness or an unavoidable catastrophe. Clearly in a real estate context, a major exception to this is in the case of leasehold investments which are effectively wasting assets; at some point in the future these interests will come to an end and will therefore have no capital value. In such cases it is necessary for investors to ensure that they receive sufficient returns to compensate for the ultimate loss of capital; additionally, part of these returns might contribute to a sinking fund, set up to replace the capital employed.

Security of capital means more than just maintaining the original investment, it also means the protection of the purchasing power of capital. A partial loss of the capital, through the reduction in the real value of the portfolio due to inflationary effects, is not an uncommon phenomenon. Investment portfolios containing a significant proportion of government securities are vulnerable to long-term interest rate fluctuations and the inflationary decline of money values. It has been suggested that common stocks tend to increase in value as money values are deteriorating in real terms, and therefore can be regarded as a reasonable hedge against inflation. However, common stocks are more risky in all other respects than fixed income securities, therefore the investor must try to strike a balance between the dangers of inflationary effects and other risks.

Security of income

Income is derived from the investment portfolio in the form of dividends (from shareholdings) and interest earned (on bonds and other debt instruments), or in the case of real estate, rent. Dividend income tends to be more erratic than interest income. The incomes derived are either used for consumption or reinvested. The planning of investment strategies and the articulation of objectives is much easier with stable income streams.

When the income is mainly used for consumption, the importance of maintaining the real value of future incomes becomes an important consideration. When assessing the long-term income needs it is important to produce a buffer against inflation. This usually means an increase in the capital commitment, which is the only reasonably certain way to ensure the production of the required income in the long term. In the early stages following initial capital investment the diminishing real value of income may not be fully appreciated. It is therefore important that care should be taken at this time to regularly reinvest suitable proportions of any income received in order to provide for long term maintenance and protection of the real value of the capital invested, and thereby the ongoing maintenance in real terms of investment income.

The main objective of investment portfolios is generally the provision of satisfactory levels of income, the achievement of capital gains in the future, or a combination of the two. Accordingly, there is a distinction between an income portfolio and a growth portfolio. The strategies appropriate for these differing objectives are significantly different, and therefore it is important to define clearly the objectives.

Marketability and liquidity

Marketability refers to whether a particular asset can be bought or sold easily, while liquidity usually means the ease with which a particular asset can be turned into cash. Both these aspects are important to those investors who may wish to take advantage of attractive opportunities that may arise. Investors often attempt to acquire stocks when the prices are low and sell when prices are high. Purchasing stock at the opportune moment may require the quick conversion of some of the existing assets into cash or cash equivalent.

Taxation

A very important factor in the selection and management of investment portfolios is the tax position of the investor. Most investment decisions are made by considering the expected returns in the light of the investor's tax liabilities.

Investment returns are generally liable to income tax, corporation tax and capital gains tax, dependant on the particular circumstances of the investor. The tax status of individuals, corporations, and institutional investors varies considerably, and the complex nature of tax planning and calculations usually necessitates that investors take specialist advice. Attempts to reduce the burden and impact of taxation have considerable implications on the setting of portfolio objectives and the planning of investment strategies. Investors who are liable to high rates of income tax must consider very carefully the various options available to them to reduce or defer the tax burden on their incomes. Of particular interest to such investors are those investment vehicles which offer some tax relief or where taxes have already been deducted at source. Some investors may prefer investments that enable tax to be deferred, and opt for investments whose returns are taxed on capital gains, rather than regular income. As a capital gain is made when the investment is disposed, the capital gains tax payment will be due after disposal; it may be possible to reduce future capital gains taxes by utilising the annual exemption allowance. In some situations investors may incur double taxation of returns, for example when corporation tax is levied on capital gains at the corporate level, and income tax is subsequently paid by individual shareholders. Tax transparent investments are investments that avoid double taxation, and there continues to be a strong call by investors for the further development of such investment vehicle for real estate. This is one of a number of aspects considered in the consultation document published by HM Treasury (2004) that looks at the issues pertaining to indirect real estate investment.

Real estate portfolio management

The maintenance of investment policy requires careful and expert management; portfolio management activity has been defined as:

> A continuous process of reviewing the portfolio to determine areas where action can be taken with a view to improving the return from an investment.

> A restless activity involving:
>
> (a) Analysing each investment property and comparing its actual performance against the expectations on acquisition and its comparison with other forms of investment in the portfolio;
> (b) Seeking ways and means of improving the performance of a particular investment;
> (c) Disposing of investments where long-term prospects are likely to be less than expected.
>
> [Jenkins, H, in Stapleton (1994)]

Although an ageing quotation, it seems to have survived the test of time, and remains as valid today as it was when it was first expressed.

The portfolio management process may be viewed as the continuing interaction of the decisions made on the basis of the outcome of portfolio analysis, and the results and consequences of the actions carried out under the heading of portfolio rationalisation or adjustment. The process is a complex one, containing a number of important and sensitive problem areas. The success of investment strategy and policy depends entirely on the thorough understanding and control of the portfolio management process. This process contains the following principal components:

(i) determination of the optimal portfolio composition
(ii) portfolio analysis comprising the retrospective analysis of performance and the prospective assessment of future performance in terms of returns and risk
(iii) portfolio rationalisation and adjustment based on the conclusions of portfolio analysis
(iv) review of the strategic and tactical investment objectives and decision criteria.

Decision making and internal and external portfolio activity are constrained and prescribed by an investor's overall investment strategy, the amount of funds available for investment in real estate during any portfolio period, and the length of a portfolio period and the investor's time horizon.

Once assembled, real estate portfolios should be managed by appropriately qualified experts who thoroughly understand the characteristics of the assets in their care, and who are equipped with the skills, knowledge and understanding to actively manage the portfolio at the operational level. Furthermore, real estate portfolio managers should also understand the principles and rationale of modern portfolio management practice, to enable them to manage the investment portfolio at the strategic level, and to ensure investment efficiency and effectiveness.

Regarding the construction and ongoing management of the portfolio and the challenges and burdens of investment decision making, the investor has a number of options. One option is for the investor to retain full control of the investments and make all the selection and allocation decisions personally. While this may be the cheapest option in terms of savings in fees payable to professional investment managers, it requires a considerable amount of time, knowledge and expertise to make it a successful and effective option. A compromise solution might be for the investor to retain the more strategic level management functions, while delegating the responsibilities of the day to day acquisition and disposal decisions to retained professionals. Alternatively, the investor may decide to engage professionals to manage all aspects of the investment portfolio, from investment asset management to the daily operational management. Specialists employed for these purposes might be expected to include members of the Financial Intermediaries Managers and Brokers Regulatory Association (FIMBRA), and the Royal Institution of Chartered Surveyors (RICS).

Large corporate and institutional investors usually have specialist in-house investment departments; nevertheless, they may also use the services of external experts to enhance the efficiency and reliability of their in-house team. Outsourcing of activities is a convenient way to manage excessive workloads in peak times, while retaining a strong nucleus of in-house staff sufficient to cope with normal work levels. Additionally it is a way of gaining access to specialist skills and knowledge without having to expand the in-house team.

In the case of a pension fund, it is essential to achieve a certain rate of return on the invested funds, in the long run, if the future benefits of the contributors are to be guaranteed. The investor is likely to require access to some or all of the capital invested, at some point in time. As far as the individual investor is concerned, the length of time that his capital is tied up in investments is an important consideration. As far as large corporate and institutional investors are concerned, they need to determine the dates at which cashflow will be required. It is unlikely that these investors will need to realise all their invested funds, nevertheless some of the foreseeable liabilities may be considerable. For example, if an insurance company is selling endowment policies which will mature in 10 and 20 years time, they will need to consider the purchase of securities which will match the expected liabilities.

Since the late 1970s there has been some evidence of a more rapid ageing process of investment properties with technical obsolescence, resulting in social obsolescence, leading to economic obsolescence. This raises the problem of deciding when, and on what scale, refurbishment or redevelopment should take place.

The investment and portfolio objectives are achieved through pursuance of appropriate policies and strategies; these entail acquisition, management and disposal of individual investments. Irrespective of the balance between retail, office and industrial property, the investment quality of the assets may vary. As far as the investment portfolio strategy is concerned, it can be subdivided into three well-defined areas:

(i) *selection*: This involves the selection of a particular type of portfolio which appears to be most appropriate to achieve the investor's objectives, and the selection of individual assets for a particular portfolio
(ii) *allocation*: To decide the appropriate level of capital commitment to the portfolio as a whole and to sectors and individual assets in the portfolio
(iii) *timing*: For acquisition, disposal and restructuring of the portfolio and its components.

The quantification of risk is usually achieved through the use of the relatively simple statistical measures of variability, such as the variance and standard deviation of expected returns; these measure the variation of an asset's returns. Movement of an asset's returns in relation to the general market's returns is referred to as volatility; volatility is a relative measure of risk of the asset in relation to the market and is usually expressed by the beta coefficient. The beta coefficient may be obtained through regression analysis of the time series of returns of the asset against the comparable time series of returns for the market; essentially the beta coefficient is the slope or gradient of the line of best when asset returns are plotted against market returns. Quantitative measures of risk are essential for analytical purposes and for the setting of the limits of risk bearing. Unfortunately, these quantified expressions of risk cannot cover the whole risk spectrum and they are imperfect, hence most practising investors tend to treat them with some skepticism and caution.

The principal objective of rational portfolio investment, to maximise returns for a given level of risk, is generally pursued through a strategy of diversification. In order to put into effect a worthwhile diversification strategy the investor needs to be in an appropriately strong financial position, which is usually the privilege of large corporate and institutional investors. These investors generally diversify their portfolios on three levels:

(i) diversification among different investment media
(ii) diversification among the different sectors of the investment media
(iii) diversification among individual assets and securities within particular sectors of particular investment media.

It has been suggested that in general, real estate portfolios tend to be poorly diversified, one measure of portfolio diversification being the coefficient of determination (R^2). This statistical measure can be used to measure the proportion of variance in the returns of a portfolio which is explained by the performance of the overall market portfolio. R^2 can take values within the range 0 to +1; the closer it is to +1, the greater the percentage of the variation in the portfolio's return from its expected return, which is explained by the performance of the market portfolio. It has been suggested that a well diversified portfolio should have a R^2 value of at least 0.95, which would require an equally weighted portfolio of at least 200 properties. This is a considerable size for a real estate portfolio which only the larger institutions would be able to hold.

There are two approaches to the planning of portfolio strategies. One is usually referred to as the traditional approach, which simply looks at the investor's objectives in terms of the need for income or capital appreciation and then selects those securities which appear to be the most appropriate to meet these needs. The other approach is a more theoretical one based on creating a strategy which will maximise the expected returns on a portfolio for a particular specified level of risk; this approach is based on Modern Portfolio Theory. The traditional approach to strategic planning remains well respected in spite of its limitations and idiosyncrasies, while the Modern Portfolio Theory based approach continues to be regarded cautiously, largely because of its heavy reliance on complex mathematics, and the underlying theoretical assumptions which do not accurately reflect the real world. Additionally the inherent characteristics of real estate make adoption of this approach problematical.

Measurement and analysis of investment performance

Investment performance, its measurement and analysis, have attracted considerable attention since the 1960s and early 1970s. Since then a whole industry has developed, offering professional advice concerning the measurement of investment performance and the expert management of investment assets. In the early days of this development most of the activity was concentrated on the performance measurement of investment portfolios containing stock market securities. In more recent times attention has been given to the need for reliable measurement of the investment performance of real estate assets.

The large investment portfolios of various mutual funds, pension funds, insurance companies and other financial institutions, contain substantial holdings of assets other than stock market securities, such as real estate and works of art, etc. The trustees of such large funds realise the need for the measurement of historic performance and the assessment of the likely future performance of their investment portfolios.

As previously mentioned, performance may be defined as achievement relative to targets and objectives, by measuring performance, the degree of achievement against a set of objectives and targets can be expressed in quantitative terms. The shortfall or excess, relative to the targets, can then be analysed and useful conclusions and explanations drawn for decision-making. Through the monitoring and the analysis of portfolio performance, investors can gain valuable insights into the investment characteristics and behaviour of the various assets included in their portfolios. Such characteristics and behaviour can then be considered against the context of the movements of the various sectors of the investment market. Only through a standard method of performance measurement can the true contribution, good or bad, of the different assets, groups of assets, and sectors of the portfolio, to the general investment portfolio, be evaluated.

In the early stages of the development of this process, the measurement and assessment of investment performance was regarded only from the point of view of the investor or the trustees. However, it is now recognised that fund and portfolio managers also require the benefits of performance measurement and analysis for internal, operational use; at the operational level, the benefits of performance analysis are considerable. Performance analysis is an extremely valuable analytical decision-making tool for the monitoring of the achievement of the targets set by the investor or trustees for the portfolio manager.

Investment portfolios are constructed and managed to meet the varying requirements and parameters of investors. Differences between income and liabilities, the relative attractiveness of income or capital growth, and the alternative mediums for investment, make the work of analysts complicated, and the results of analyses liable to criticism on various grounds.

In the late 1960s and early 1970s, studies by the Bank Administration Institute (BAI) (1968) in the USA, and the Society of Investment Analysts (SIA) (1972) in the UK, laid the foundations of a more standardised approach to the investment performance measurement of gilt and equity portfolios of financial institutions. Recommended measures of total returns were the time weighted rate of return by the BAI, and the Money Weighted Rate of Return by the SIA; for risk it was suggested that the variability of total returns could be used. At that time, measurement and analysis of real estate investment performance was largely undeveloped, and did not fall within the scope of the recommendations of either the BAI or SIA, albeit the difficulty of undertaking analysis of real estate investments due to the absence of a reliable market index was acknowledged.

However, during the latter part of the 1970s and throughout the 1980s, considerable developments occurred within the context of real estate investment performance measurement and analysis. In the UK property consultants and advisors began to offer portfolio performance measurement and analysis services to investors, and the report of the Wilson Committee in 1980, which looked at issues relating to the accountability of pension funds, lent further support to this developing field of inquiry.

Real estate investment assets and portfolios present great complications if they are included in the measurement of performance of a mixed asset investment portfolio. While the market prices of the stock market securities are available instantaneously on an objective and consistent basis, the market price of real estate assets are subjectively estimated by using some appropriate valuation method. These complications are such that it is best to examine real estate assets separately in the context of specialised real estate portfolios and then regard the real estate portfolio as a single asset in the global investment portfolio.

Today performance measurement and analysis of real estate investments is well established, although theoretical and practical issues and difficulties remain unresolved. For example the particular difficulty of measuring capital returns due to the reliance on valuations (an estimation of price), rather than the actual observed transaction price, and the use of statistical measures of variance and standard deviation as a proxy for investment risk.

Institutional investors and trustees require performance measurement and analyses for the following reasons.

- *To monitor the progress of invested capital*: This monitoring is absolutely essential as investments are usually held to meet certain liabilities in the future. Trustees and managers must ensure that the progress of investment assets enables these liabilities to be met when they fall due.
- *To monitor relative performance*: Institutional investors like to pitch their expertise against the expertise of the managers of other funds, they are particularly interested in how well, or how badly, their investment activities fared in this respect. Portfolio performance can be compared with the performance of investment markets, and also against the ups and downs of the economy, as reflected in the general economic indicators. Additionally a portfolio's performance may be compared more directly against those of competitors, through benchmarking and performance indicators and league tables.
- *To analyse past performance for decision-making*: The allocation of funds to new investment activity may be informed by the analysis of past performance of portfolio structure and stock selection. Historic overall portfolio returns may be assessed in terms of both the effect of portfolio structure (and therefore investment strategy), and the effect of the performance of stock held within the portfolio. From the analysis, the relative weight of strategic elements, ie the proportions held in various asset groups and sectors, and the effectiveness of stock selection may be identified. From this analysis, decisions regarding the correction of imbalances can then be made.
- *To evaluate the performance of management*: Performance measurement will provide an insight into the effectiveness and efficiency of the investment managers. This is particularly important if the responsibility of fund management is outsourced to external investment managers and analysts, but is equally as important from the point of view of individual investors who want to know how well those charged with the management of their investment funds have performed.

- *Operational decision-making and reflection*: At the operational level, analytical tools are needed which can be used by the portfolio manager to evaluate the performance of the investment assets in the care of the manager, and to facilitate analysis of the effectiveness of the manager's previous decision-making; the results of this analysis will inform the ongoing operational decision-making.

The measures and indicators of performance are expected to express in quantitative terms, the achievement of targets and objectives in the context of the market which generates the risk return relationships or trade offs. The rate of return is regarded as a key measure of performance; the time weighted rate of return and money weighted rate of return being commonly used for this purpose.

The performance of a portfolio or fund is usually interpreted in a broader sense than the rate of return; another important element of performance is the risk associated with the portfolio or the fund. The different aspects of risk can be reflected and assessed in terms of the following:

- the variability of the rate of return on individual investments, groups of investments or sectors of the portfolio, and the portfolio as a whole
- the volatility of the rate of return on the portfolio which indicates its sensitivity to the rate of return on the market
- the diversification of the portfolio expressed by the portfolio balance
- the downside risk of the portfolio, which is the probability that a specified target rate of return will not be achieved.

The use of indices in real estate investment analysis began to receive closer attention in the early 1970s. Indices were seen as promising tools for revaluations; their use for performance measurement and analysis being suggested later. Since then the availability and quality of real estate indices has developed substantially, and indices are now used extensively for many purposes, for example:

- comparing the investment market performance of the real estate asset class with other investment media
- comparing the investment market performance of the various market sectors of real estate
- comparing the investment performance of individual real estate portfolios against aggregate market performance

- comparing the investment performance of individual properties, and segments or sectors of portfolios.

While internal comparisons within a portfolio do not require the availability of an authoritative real estate market index, external comparisons can only be made if a suitable and reliable market index or benchmark available.

If after portfolio investment performance measurement has been undertaken it is found that the original investment objectives and targets set have not been achieved, this does not automatically imply a failing on the part of those responsible for their achievement. Unforeseeable changes in conditions, lack of resources, or doubts as to the quality of the system of measurement should all be carefully considered. Further analysis might reveal that despite not achieving targets, the custodians of the portfolio have actually performed well under the particular circumstances.

Although measurement of investment performance may provide an insight into the performance of managers and perhaps even trustees, it would be wrong to automatically link the two. For example, a property manager might hold back from the market, not using the funds allocated, and permitting money to temporarily be used in another sector of the portfolio, to better effect. In addition, on occasions, trustees might delay or frustrate a particular property proposal.

Complexities, problems and difficulties

The assumptions underlying portfolio performance measurement can be subject to a number of criticisms. The extent to which valuations are proved in the market place is the most serious, since if valuations are not accurate, conclusions based upon quarterly, half-yearly or annual analysis will not command confidence or respect. While studies have suggested that valuations may generally provide a reliable proxy for market prices, the accuracy of valuations remains a fundamental and unresolved issue for real estate investment performance measurement; unless capital value data is reliable and accurate, much of the work on measuring return and risk is fundamentally flawed.

The relevance of the classification of real estate into categories or sectors for the purpose of analysis is questionable in cases where the actual use does not meet the general expectation for the particular property type. For example, many high technology companies occupy industrial space, yet use it as if it were predominantly office space. For instance, an information technology based company taking accommodation on an industrial park may require 25% offices, 15% testing or research space (which looks like offices), 30% light assembly, and 20% storage (with the remaining 10% for circulation and staff facilities). Such a unit is much more akin to the offices sector, though located in an industrial area. Additionally the development and growth of out of town retailing, retail parks, and retail warehouses add further complexity to the classification of retail space; the growth of business parks in contrast to traditional city offices makes the classification of business space equally challenging. In all, the developing nature of space and space requirements presents ambiguity, and a potential dilemma when attempting to categorise by property type. In 2005 the reclassification of retail use classes will further affect the situation.

Different investors have different requirements and liabilities and so need to pursue different investment polices, even if their real estate portfolio type and sector balances appear similar to those of the market index and norms. Although in cases where the subject portfolio has a different structure and profile to that of the market index, and therefore comparison against the index at the portfolio level may not be appropriate, sub sector analysis may still be possible. Additionally it may be possible to construct an overall portfolio benchmark that reflects the property type balance of the subject portfolio.

Whatever the merits, mechanism and historical conclusions which can be drawn from a study of past performance, the fundamental reason for making an investment is a future expectation of an income flow. Hence, real estate performance measurement and analysis leads inevitably to the estimation of future changes in the factors influencing values; these fall into two categories:

- national macro-economic factor, and
- local factors related to individual properties and their environs.

Study of the former is relatively well advanced in general economic models, this being reflected in the relative success of some retail and industrial rental forecast models which appear capable of successfully projecting rental levels in the short term.

While short term supply aspects in the form of land and property availability registers are widely reported by local authorities and private practices, the quantitative analysis of factors influencing values locally is still in its relative infancy. However, the retail sector has perhaps enjoyed greater progress than other sectors in this regard, due to the significance of population, social composition, and earning statistics. Research departments in real estate practices, and within financial institutions, continue to seek to develop analytical bases and models with which to accurately estimate future changes in real estate values.

Market indices and benchmarks

Investors, large and small, corporate and private, recognise the need to examine the performance of their investments and investment portfolios in the context of the wider investment market; they are generally concerned about their own performance and the performance of their portfolios relative to the performance of others.

This requires that there is a suitable yardstick or benchmark against which the performance of individual investments and portfolios may be judged, thus enabling it to be determined whether the portfolio has achieved an above or below average measure of success. As investment performance will be judged against an appropriate market index and possibly a fund's or portfolio's position in a league table, this perhaps explains the importance of index tracking as an investment strategy, as well as the desire of portfolio managers to outperform the index.

There are a number of indices in general use in investment practice. Most of these indices reflect market movements, and the various stock markets and security markets; some of them, for example the Retail Price Index, are used as general indicators of the state of the economy.

In the stock markets, the use of indices in investment work and performance measurement is widespread. Various gilt, share and bond indices provide insights into the state of the whole market, or into the level of activity in the sectors of these markets. While these markets are well provided with sophisticated and reliable market indices, other sectors of the investment spectrum are less well catered for in this respect, the real estate market being one such example.

Historically the real estate investment market was ill equipped with reliable market indices, although since the 1980s this situation has changed, and now there are a number of respected indices available, the Investment Property Databank (IPD) indices being perhaps the most widely known source of real estate indices in the UK. The main reasons for the slow evolution and development of reliable real estate market indices were the fragmented and localised nature of the real estate market in which individual investments are unique, the relatively low turnover of stock, the inherent secrecy within the marketplace, and the inherent problems of obtaining data, let alone reliable data. Although the reliability of property indices has improved considerably, these features of the property industry remain, and continue to affect the process of index construction.

A further issue relating to the construction of property market indices is the potential for bias in the event that capital and rental valuations that contribute to the index are carried out by a relatively small range of firms. The RICS Appraisal and Valuation Standards and the Carsberg Report set out changes for major appointments. In an attempt to safeguard against this it has been suggested that the following procedures should be adopted:

- the database should be sufficiently large to enable reliable samples to be drawn from it
- data should be obtained from a representative range of competent valuers
- the range of data about each property should be sufficiently wide
- the independence of the sources of data should be available for verification.

There are a variety of property market indices and indicators, covering both rental and capital value changes. These indices may be classified as follows:

- whole fund indices
- indices derived from special index portfolios
- indices based on data drawn from special locations.

Most of the indices are desegregated into:

- capital or capital growth indices
- rental or rental growth indices
- total return indices.

Whole fund indices are constructed from the data and information drawn from all the subscribers to a particular service. The subscribing funds are combined into an aggregate fund. The data collected refers to the participating fund as a whole, rather than data pertaining to individual properties in the investment portfolios. Such indices are based on data on the performance of actual funds which are willing to provide information.

Typical indices that might be produced include: an index of property values (capital value index), an index of total return, and an index of income return. Such indices can be produced on the basis of information covering the following:

- the open-market values at the beginning of a period
- purchases and sales made during the period
- net rental income during the period
- the open-market values at the end of the period.

The actual calculations of the indices involve linking the periodic figures; the weighting being undertaken on the basis of capital value.

The second category of indices is those derived from special index portfolios or data on individual properties aggregated into representative portfolios. The producers of these indices are either large surveying firms who manage a substantial proportion of the stock of investment properties of institutional investors, or an organisation sponsored by a substantial part of the representatives of the property investment market. These producers have access to substantial amounts of pertinent data so that they can create notional market portfolios by aggregating selected properties owned by their clients, into large representative portfolios. As a consequence, they are able to create a model of the institutional property investment market. The difference between this approach and the whole fund approach is that the data is drawn from the individual properties included in the notional portfolio. The notional portfolio is then used to provide the data for the construction of the index.

Although the indices provided by two of the main providers are of the same general type, they have significant differences. Perhaps the most significant difference is the way in which the indices are produced. One of these providers is an independent research company that collates data for the construction of indices from over 100 different firms of surveyors. The other is a company of real estate advisors who, like a number of other similar companies, produce indices using data from the company's own sources, relying on their own in-house valuation and management statistics.

The third category of index, indices based on data drawn from special locations, reflect the changes in rental levels. They do not refer to specific properties but to "rent points", selected in the regions of the country, from which samples of rents are collected and weighted to reflect an institutional portfolio. There are some variations in the manner these indices are constructed, but the objective is to trace the market movements of rental incomes in the context of institutional investment activity.

This brief review of the three main categories of property indices reveals that there are variations in concept, style and format of the various indices. Although there are differences between them, they all attempt to reflect the state and trends of the property market. Since the 1980s, the indices have become more sophisticated and more reliable; great care has been taken to use a wide spread of data to represent all major sectors of the property market. The data collection and the quality of data, and the management and maintenance of databanks, are the main contributors to the increasing reliability of property indices.

There are still issues concerning the limitations of property indices, not least the mathematical and technical impossibility of producing a single index of property performance covering the full property spectrum. However, indices are now more highly regarded across the property industry, with sector indices like IPD's commercial property index being widely used as a benchmark of the sector's performance.

The construction of real estate market indices is complicated by the nature, availability and collection of data; the data requirements are such that specialised data banks are required. Without an efficient and well designed data bank the provision of a reliable real estate index is virtually impossible. In recent years IPD has managed to establish such a data bank, drawing real estate data from many of the major financial institutions and investor, with which they are able to construct a wide range of real estate indices. These indices are regarded by many as the market leaders, and command widespread respect and confidence in the market place. However, in view of the highly fragmented and diverse nature of the real estate market, which consists of numerous local and specialised markets, an overall real estate index is always likely to remain illusive, or at best have inherent limitations. Notwithstanding this point, such general indices do provide a valuable insight into the generality of market performance.

In addition to comparing the performance of individual, sectors, and whole portfolios of real estate assets against appropriate market indices, performance may also be undertaken by reference to league tables. These are produced by analysts to rank the performance of investment funds or portfolios. However, as league tables may be constructed without full regard to the objectives of the fund or its particular characteristics and level of risk, interpretations and conclusions drawn from these tables should be regarded cautiously.

Operational or corporate real estate portfolios

As previously mentioned in this chapter, substantial property is held by individuals, companies and corporate organisations, public and private, not specifically for investment purposes, but for operational purposes. Operational or corporate real estate is effectively property that is held as a factor of production by parties ostensibly not in the specific real estate business. The remainder of this chapter examines issues related to the management and performance of corporate real estate portfolios.

Corporate real estate and its management

Many organisations are involved with real estate as a product in its own right, for example investors including financial institutions and real estate development companies. However, in addition to requiring real estate for investment purposes, it is also required and regarded as a factor of production, such real estate being commonly referred to as corporate (or operational) real estate. Corporate real estate is defined as real estate held for operational purposes, ancillary to the main purpose of a business or organisation.

Together with capital and labour, real estate is traditionally regarded as one of the three principal and essential factors of production contributing to the production of commodities. Arguably, all businesses and service providers require some form of real estate base from which to operate, and therefore it may be considered a prerequisite for businesses and service providers alike, and an essential factor underpinning all economic activity. However, as noted by Weatherhead (1997), following the emergence of cyberspace and the virtual office, some businesses have configured themselves to operate with very modest real estate requirements.

Given the general significance of real estate to all kinds of organisations, and regardless of the amount employed, careful and efficient management of this important, limited, and costly resource, has become a logical and important objective for most owners, occupiers and users.

The management of real estate has been an activity since time immemorial, a professional occupation of longstanding, and the subject of ongoing study and research. Much of the research focus has been on the management of real estate held for investment purposes; prior to the 1980s relatively little attention had been paid to the specific issues concerning corporate real estate.

Working at Harvard University in the USA, Sally Zeckhauser and Robert Silverman (Zeckhauser and Silverman 1981 and 1983) drew attention to the significance of corporate real estate, observing that the management of this important resource was being widely neglected by corporate organisations. This seminal work is widely acknowledged as a principal factor in raising the profile of corporate real estate and its management. Since then, corporate real estate and corporate real estate management have become the specific focus of growing academic and industry based research interest.

In the UK, interest in corporate real estate also stirred in the 1980s and resulted in a number of government agency studies into the management of public sector real estate, for example the work of the Department of Health and Social Security (1982); Cabinet Office (1985); Audit Commission for Local Authorities in England and Wales (1987a, 1987b); National Audit Office (1988a, 1988b, 1989, 1994, and 1996). Such studies have focused on general property management issues as well as specific categories of real estate, for example police and educational real estate.

In the latter part of the 1980s the strategic management of corporate real estate became the focus for attention in the UK, resulting in the publication of the findings of the first major study in the UK into the state and nature of management of both public and private sector operational property assets (Avis, Gibson and Watts, 1989). This study specifically highlighted the importance of real estate as a valuable resource, and the need for its effective and efficient management; the study has subsequently been developed and updated (Avis and Gibson, 1995).

Since then there have been a range of academic and professional studies into aspects of this topic area, resulting in a considerable expansion of the literature, both in the USA and UK, and increasingly in respect of corporate real estate in other countries.

It is perhaps surprising that interest in the strategic aspect of real estate did not occur earlier, given that general literature relating to corporate strategy in business dates back at least to the early 1970s, and possibly even earlier. Of equal surprise is that general business literature, which includes the important and well known contributions by Porter (1980), Hamel and Prahalad (1994), and Collis and Montgomery (1995), tend to pay little attention to the role and contribution of real estate in achieving corporate strategy. It was only in the 1990s that the strategic role and contribution of corporate real estate to the overall success of organisations really began to be addressed by real estate practitioners and researchers.

This absence, until comparatively recently, of both business and real estate literature focused on the strategic management of corporate real estate, has been mirrored in organisations holding corporate real estate. While some organisations now have greater regard for the effects of real estate on corporate strategy and overall business success, and take steps to incorporate real estate strategy into corporate strategy, many organisations are still believed to rank real estate low in importance in comparison with other aspects of the business. It has now been recognised that failure to take account of the full role and potential contribution of real estate to organisations, and the continued neglect of this resource, may work to the detriment of organisations. It is this realisation that has led to the contribution of corporate real estate to the overall success of organisations, becoming the specific focus of research by real estate analysts, strategists and advisors.

It has been suggested that the neglect of real estate may be explained by the general absence of property professionals in sufficiently senior positions within organisations to manage and advise on real estate issues at the corporate level, and thereby to raise the profile of real estate and help ensure its role within the organisation is understood. If the role and contribution of real estate within organisations is to be appreciated, then it needs to be represented at the highest level by appropriately qualified personnel, and the real estate function given its due status.

Research undertaken since the 1990s into the contribution and effects of corporate real estate has revealed some stark relationships. For example, it has been suggested that in large UK businesses, on average, real estate accounts for between 30-40% of asset value, and between 16-17% of annual total costs. Citing the work of Bannock and Partners (1994), Weatherhead (1997) comments that if real estate accounts for 20% of annual costs, a 5% saving in real estate costs will result in a 1% cost saving overall. Depending on the gross operating margin (relationship between total costs and turnover), the effect on operating profit may be significant. Consequently, it is clear that real estate plays an important part in overall business success and therefore should be more fully considered when determining corporate strategy.

Analysis of corporate real estate literature reveals a broad range of issues that have been at the centre of corporate real estate management related research and discussion. Some of these issues are focused on the nature and characteristics of corporate real estate itself, others relate more directly to the process and state of management of this resource. Manning and Roulac (1999) and Roulac and Manning (1999) provide an overview of corporate real estate related research, and observe how interest has developed, and how since the 1980s much of the research has been driven more by industry than by academics. However, despite the attention that the general topic of the strategic management of corporate real estate is now receiving, research in this area is still relatively immature.

The relationship between corporate real estate management and facilities management

Corporate real estate was referred to by Zeckhauser and Silverman (1981 and 1983) as "... the land and buildings owned by companies not primarily in the real estate business." Companies in the real estate business are generally taken to include property developers, investors and traders. This definition seems to have gained widespread acceptance; it being adopted by many subsequent writers on the subject. Corporate real estate may include a wide range of property types or categories for example: retail, office, industrial, leisure, public and private; the form or title of ownership may be freehold or leasehold. However, whilst a company may not primarily be in the real estate business, it is interesting to note the assertion by Zeckhauser and Silverman that "... a $1 billion company is in the real estate business, probably to the tune of $250 million". The inference being that many non real estate specific companies (if not all), are in fact in the real estate business as a consequence of their investment in this resource.

The basic concepts of corporate real estate management have been outlined by Bon (1992); the following quotation succinctly summarising this activity as follows:

> CREM [Corporate Real Estate Management] concerns the management of buildings and parcels of land at the disposal of private and public

> organisations which are not primarily in the real estate business. ... CREM covers the entire range of activities concerning portfolios of buildings and land holdings: investment planning and management, financial planning and management, construction planning and management and facilities planning and management ... The primary concern of CREM is to establish and maintain a close match between an organisation's business and property strategies.

Corporate real estate management includes a diverse range of concerns and activities; the key concern perhaps being the strategic management of this resource and the need to align property strategy with overall business strategy. This view has been central to the development of corporate real estate management as an activity, in that strategic management of corporate real estate provides considerable opportunity to make important and substantial contributions at the corporate level.

Since the 1980s, facilities management has grown alongside corporate real estate management as both a function and a professional activity. There now exists extensive literature on this topic that covers issues ranging from the very technical to the strategic, and the nature and development of the facilities management role. Definitions and interpretations of the function of facilities management abound, but the following definition by Alexander (1992) is typical:

> ... facilities management is concerned with the integration of property management, with the management of its utilisation and with the full range of services provided to support a business operation. As such it will operate at three interrelated levels in any organisation — strategic, tactical and operational.

Kincaid (1994) comments that facilities management is the integration of property (real estate) management, property maintenance and operations, and office administration. Alexander (1994) focusing specifically on strategic issues relating to the development of facilities management as a profession summarises the facilities management movement as:

> ... a belief in potential to improve processes by which workplaces can be managed to inspire people to give of their best, to support their effectiveness and ultimately to make a positive contribution to economic growth and organisational success.

Significance of corporate real estate — key observations

Criteria	*Observation*	*References*
Proportion of a company's total assets accounted for by corporate real estate	A range of figures have been suggested, but these generally lie in the range 10–50%	Veale, 1988. Zeckhauser and Silverman, 1981, 1983 Avis, Gibson and Watts, 1989 Bon and Luck, 1999
Proportion of operating costs accounted for by corporate real estate	A range of figures have been suggested, but these generally lie in the range 10–50%	Veale, 1988 Zeckhauser and Silverman, 1981, 1983 Bon and Luck, 1999.
Property occupancy costs as a proportion of pre-tax gross sales	5–8%	Veale, 1988 Avis, Gibson and Watts, 1989
Property occupancy costs as a proportion of net income.	40-50% of net income	Veale, 1988
Contribution to Gross Domestic Produce.	10%	Debenham Tewson Research, 1992
Real estate operating costs	Second only to payroll costs	Veale, 1988

While not exhaustive, the above definitions and comments provide insights into the nature of corporate real estate management and facilities management and their similarities and overlap. The overlapping definitions are perhaps an indication of the competitive rivalry between the facilities management and property management professions, both of whom seek to include the strategic management of real estate within their remit. Whether corporate real estate management is a subset of facilities management, or whether facilities management is a subset of corporate real estate management, is a matter for ongoing professional argument and debate. In so far as the remainder of this chapter is concerned, the position taken on this matter is that facilities management is largely (although not exclusively), embraced within corporate real estate management.

Significance of corporate real estate to organisations

One of the underlying reasons for the growth in interest in corporate real estate and its management was the recognition that it is of considerable physical and financial significance to companies and organisations. A number of studies have investigated the extent of this significance by examining descriptive aspects such as floor area, cost and/or value, proportion of total and/or fixed assets, the proportion of annual expenditure spent on operating costs, and the effects of savings in operating costs on profitability; many other studies into corporate real estate and its management have drawn upon and accepted these observations. Some key observations that emerge from these studies are summarised in the table on p304.

The observations and studies cited above relate to both private and public sector corporate real estate. The studies illustrate the extent of the financial commitment that corporate real estate represents to organisations whose primary business is not real estate. Whilst there is variance and uncertainty in the actual figures for corporate real estate as a proportion of total asset value, or occupancy costs as a proportion of corporate income, sales, net profit etc., the underlying message is clear; whatever the precise amounts, the proportion is significant, making corporate real estate an important resource and concern to organisations. Consequently the observations lend weight to the view that, for the benefit of organisations and businesses, corporate real estate needs to be managed carefully, competently, effectively and efficiently.

However, in addition to the overwhelming financial and management significance of corporate real estate, Veale (1989), acknowledging the work of Becker, Margulis and Marans comments:

> ... the impact of the physical environment on organisational productivity and corporate mission, while more difficult to measure, is of equal if not greater importance than the cost or value of the real estate itself.

Thus it is not just those physical and financial factors relating to real estate that business managers need to be aware of and to manage, but also the effect that real estate has on corporate performance and success. For example, what is the effect of corporate real estate on employee motivation and production, as well on the customer? While there is growing realisation that the corporate real estate provision can affect the revenue side of a company's business, measuring this affect is not so easy as measuring asset value, running costs, floor space etc. The search for data and measures of performance to provide insight into such matters is the focus of continuing research.

Strategic nature of corporate real estate

It is an increasingly accepted view that corporate real estate and its management have contributions to make at the strategic level. This may be as a result of the wide reaching effects of many corporate real estate decisions on the business, for example decisions on issues such as real estate acquisitions, disposals, redevelopment, utilisation etc. The underlying requirement in order to achieve maximum benefit is for corporate real estate issues to be explicitly considered within corporate strategy, and for a real estate strategy to be drawn up that fully reflects corporate strategy, thus ensuring alignment of corporate and real estate goals.

A study undertaken in the USA, phase one of the Corporate Real Estate 2000 Project (Joroff, Louargand, Lambert and Becker, 1993), supports real estate's strategic role, suggesting that real estate is a major resource available to organisations, alongside capital, people technology and information." The report observes that:

> ... a corporation's real estate — its land, buildings and work environments — is a powerful resource whose strategic value is just emerging.

The underlying reasons for this growing interest in the strategic role of corporate real estate are the beliefs that corporate real estate:

- is significant and important to organisations
- decisions can make a difference
- can benefit the organisation by enhancing its performance
- can contribute to business performance and success.

To achieve these benefits undoubtedly requires that corporate real estate is competently, effectively and efficiently managed — a realisation that raises the value and importance of the corporate real estate management function. This realisation of the potential for real estate to benefit the organisation or to add value and contribute to business performance and success, has resulted in it being promoted as a strategic asset that should be managed accordingly.

Since the studies by Zeckhauser and Silverman in the USA, and Avis, Gibson and Watts in the UK, the corporate real estate profession's argument that corporate real estate is a strategic resource continues to grow. The emergence of specifically targeted academic journals and industry led or sponsored studies and reports, the formation of an international association of managers of corporate real estate (NACORE), and the introduction of postgraduate programmes of study, provide clear evidence of a growing field of study and developing area of professional activity. Surveys of the state of corporate real estate practice undertaken since the work of Zeckhauser and Silverman suggest that organisations have started to recognise real estate as a strategic resource and have begun to embrace it in strategic plans. Examples of such studies include those by Avis and Gibson (1995), Bon and Luck (1999), and Schaefers (1999). These surveys tend to indicate growth in real estate strategies within organisations, closer linkage or alignment of corporate strategy and real estate strategy, and growth in the status of corporate real estate management departments. Furthermore, studies have attempted to explore the application of strategic models from the field of business and management to real estate. For example see Nourse and Roulac (1993), Roulac (1999), and O'Mara (1999).

The promotion of real estate as a strategic resource is justified on the grounds of its capital and operating costs — as evidenced by the various studies previously cited. Corporate real estate's potential to contribute by enhancing income generation is a further aspect that supports the argument that it is a strategic resource. This latter point is an aspect that is receiving increasing research attention, with interest being expressed in the development of performance measures to assess the "quality" of buildings or real estate (Gibson, 1999).

However, in contrast to the assertion that corporate real estate management is by implication a strategic function, it is appropriate to reflect on the views expressed by Grimshaw (1999), on the closely related function of facilities management. Grimshaw questions the abundance of definitions of facilities management that emphasise the link between the physical facilities and organisational environment, suggesting that the function of facilities management is much more diverse. He observes three paradoxes relating to facilities management. First, it is claimed facilities management is strategic, when most facilities management practitioners are at the operational level in organisations. Second, facilities management purports to be at the centre of organisational development when often it is outsourced or delivered by in-house teams set up as internal consultants. Third, facilities management claims to be proactive in managing change when in reality it is usually reactive. The extent to which these paradoxes might equally apply to corporate real estate management is worthy of consideration. If they do apply, then this would undermine the assertion that a shift in management style from the traditional, passive approach, to a more strategic, proactive approach is taking place within corporate real estate management.

Not surprisingly, recognition of the financial, physical and strategic significance of corporate real estate has resulted in attention being given to the performance of the asset itself as well as the performance and nature of the management function. These are issues that are considered next.

Corporate real estate performance measurement

Investigation into the performance measurement of corporate real estate has focused on two main perspectives; the property itself, and the management function. These aspects are considered below and in the following section.

Measurement and analysis of corporate real estate performance is increasingly regarded as an essential element of the management process. However, this situation has only developed during the last 20 years; Zeckhauser and Silverman noting a lack of diagnostic tools to evaluate performance of corporate real estate. However by the end of the 1980s, Gale and Case (1989) noted that many organisations were undertaking performance measurement. Survey results produced by Gale and Case in the USA show that 88% of the respondents indicated use of return on investment as the standard used for performance evaluation, with other measures concentrating on budget based approaches. However, in the UK, Debenham Tewson Research (1992) concluded that over 75% of the organisations were unable to assess their real estate in cost effective terms, and that systems were not always in place to enable real estate to achieve the maximisation of cost efficiency to the core business.

During the 1990s interest in performance measurement in relation to business and real estate has developed considerably, with some broadening of the type of measures used, and less reliance on financial measures. As part of this development, Kaplan and Norton (1992) introduced the balanced scorecard, a framework for measuring business performance that attempts to complement:

> ... the financial measures with operational measures on customer satisfaction, internal processes, and the organisation's innovation and improvement activities — operational measures that are the drivers of future financial performance.

The scorecard is likened to "... the dials in an airplane cockpit" giving "... managers complex information at a glance". The scorecard is born out of the realisation that no single measure will do and that a more balanced range of performance information embracing financial and operational aspects is needed. Four perspectives of measures were proposed namely financial, customer, internal business, and innovation and learning; for each perspective, goals need to be established enabling a limited number of appropriate measures to be selected which relate to the specific business or organisational goals. Figure 10.1 illustrates the appearance and concept of a balanced scorecard.

The basic concept of the balanced scorecard is that it seeks to encompass and encourage the use of wider performance measures, for example measures of customer and employee satisfaction, in addition to financial measures. Further details of the application of the method were developed in Kaplan and Norton (1993), including a framework for linking the process of performance measurement to corporate strategy; this framework is summarised in Figure 10.2.

Gibson (1999) considered the potential application of the balanced scorecard framework for performance measurement in a real estate context. Modifying it to suit the particular circumstances presented by corporate real estate, and suggesting that measures relating to corporate real estate can be developed for each of the different perspectives, Gibson suggested that:

Figure 10.1 The Balanced Scorecard

Financial Perspective		**Customer Perspective**	
(How do we look to shareholders?)		*(How do customers see us?)*	
Goals	Measures	Goals	Measures
1. ...		1. ...	
2. Etc.		2. Etc.	
Internal Business Perspective		**Innovation and Learning Perspective**	
(What must we excel at?)		*(Can we continue to improve and create value?*	
Goals	Measures	Goals	Measures
1. ...		1. ...	
2. Etc.		2. Etc.	

Source: Kaplan and Norton (1992) (adapted)

Figure 10.2 The Balanced Scorecard

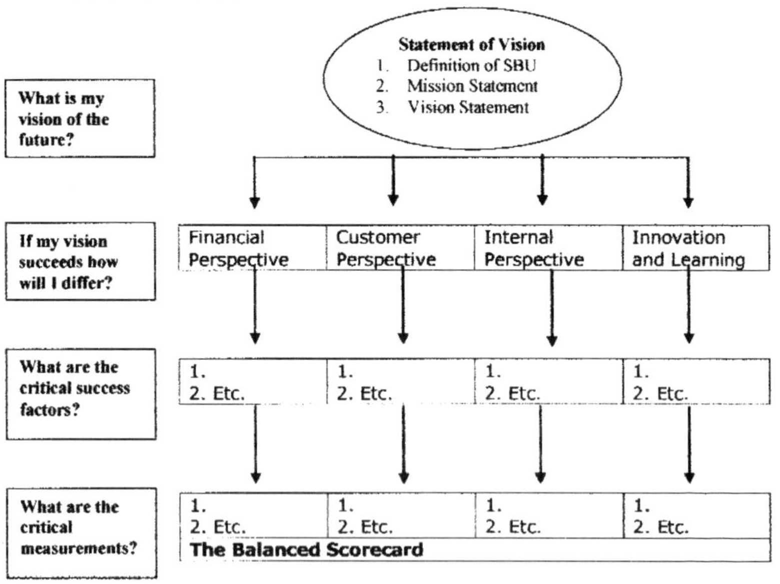

Source: Kaplan and Norton (1993) (adapted)

- financial perspective can be measured by the affordability ratio (total occupancy costs as a percentage of gross income)
- customer perspective can be measured by assessing their perception of the service or experience provided
- internal business perspective can be measured by understanding and quantifying how the physical environment influences business performance
- innovation and learning perspective can be measured by the flexibility of corporate real estate and how quickly change can be accommodated.

The four perspectives suggested within the balanced scorecard provide a comprehensive framework for performance measurement.

A research theme that has emerged in recent years is the quest for a measure of quality or effectiveness of buildings, which fits into the customer and/or internal business perspective of the balanced scorecard framework. It has been argued that quality may be a competitive weapon in so far as quality environments can be attractive to staff and customers, quality locations can improve turnover, and quality buildings can enhance profile and reputation. In attempting to measure quality or effectiveness, it is essential that objectives are clear, so that the extent, and how well objectives have been achieved can be determined. While considerable effort has gone into the development of financial and efficiency measures, which essentially involve tracking costs and the use of resources to outputs, much less effort has gone into measuring building quality or effectiveness.

Examples of efficiency measures given include costs per area and person, utilisation, and occupancy costs. The affordability ratio (total occupancy costs as a percentage of gross income) attributed to Ford (in Gibson 1999) is highlighted as a measure increasingly used by organisations.

Building quality is referred to by Gibson (1999) as "The Elusive Measure". While such a measure has yet to be adopted by managers, two contrasting approaches have been suggested. The first involves constructing a framework in which employees, managers, and customers can rate a building on a range of criteria important to them; the ratings then being used to produce a measure. The second approach is to find surrogate measures, for example staff absenteeism, recruitment or retention.

This more qualitative aspect of performance measurement of corporate real estate, concerning the measures that link building

Figure 10.3 Real Property Performance Indicator Matrox

		Category of Measure			
		Capacity	*Cost*	*Condition*	*Location*
Organisation perspective	*Employee*	Building amenities	Commute time	Work environment	Local amenities
	Facilities management	Building systems	Operating cost	Physical condition	Local infrastructure
	Real estate	Rate of return	Level of risk	Customer satisfaction	Market conditions
	Business unit	Flexibility/ utilisation	Occupancy cost	Employee satisfaction	Site proximity
	Corporate	Return on assets	Occupancy cost	Corporate image	Community relations

Source: Duckworth (1993)

performance, process, environment, property, and people, is the focus of ongoing interest and research, and will hopefully, in time, be rewarded with the development of appropriate and valid measures.

Frameworks for a balanced set of performance measures within a corporate real estate context have been devised and proposed, for example the "KPM Matrix for Corporate Property" (Proctor 1999), and the "Real Property Performance Indicator Matrix" (Duckworth, 1993).

The framework proposed by Duckworth which provides a methodology for a structured approach to the design and monitoring of a set of performance indicators for real estate, is of particular note in that it encompasses softer measures of performance, for example employee and customer satisfaction — an attempt to reflect real estate's behavioural influences. Design of the measures involves the construction of a matrix in which each column represents a category by which performance should be measured, and each row represents an organisational perspective against which each measure should be considered. Each cell in the matrix identifies the measure(s) that might be used for the particular measure of organisational perspective. As well as financial and efficiency measures, the matrix accommodates quality or effectiveness measures such as customer and employee satisfaction. The composition of the two dimensions of this matrix are summarised in Figure 10.3.

It is suggested that analysis of the portfolio of properties over a time series will provide a broad and deep understanding as a result of the synthesis of cross-sectional analysis of the real estate portfolio over a period of time.

A study commissioned in 1998 by the Higher Education Funding Council for England, Scotland and Wales (IPD Occupiers Property Databank in association with GVA Grimley, 1999), resulted in the identification of estate management statistics, definitions, key performance indicators, comparative information, and recommendations for the extension of results to the higher education sector. The study was driven by the desire of estates directors in the higher education sector to have improved data and performance measures to enable better comparisons between institutions. The resulting report proposed fourteen subject areas for performance measures, referred to as key estate management ratios (KERs), which are grouped under four main headings:

- meeting business needs
- controlling costs

- effective utilisation
- managing well.

With perhaps one exception, it seems that all of these ratios are cost and efficiency driven. Numerators of the ratios focus on cost, consumption, functional suitability, revenue, condition, space used, and staff; with the denominators focusing on space, students, staff, revenue and costs. The possible exception referred to above is the "functional suitability ratio"; this ratio is described as measuring: "... the capability of the space to support its existing function." This ratio is determined by measuring the proportion of the total gross internal floor area in each of the following categories:

- excellent
- satisfactory
- generally satisfactory (but in need of localised improvement)
- unsatisfactory
- unacceptable.

The report suggests that a higher education institution's estates department will undertake the assessment of space into its appropriate functional suitability category; initially these assessments will be largely judgmental, based on the views of the estates department and without input from other stakeholders. However, it is indicated that in due course, and despite the difficulties involved, it would be desirable for users (for example students or staff) to participate in the assessment. To achieve this will require the definitions to be developed and honed, and a consistent methodology for the categorisation of space established. Additionally it will be necessary to develop and agree a reliable measure of the functionality of space (see IPD OPD in association with GVA G, 1999).

This research highlights the interest of the higher education sector in the management of its corporate real estate resource, and in particular the sector's wish to manage its real estate more effectively and efficiently. A consequence of this may be a prevailing management approach to corporate real estate in higher education that is proactive and strategically oriented. While of specific relevance and application to the higher education sector, the recommendations of the report, for example the key estate management ratios, have potential to be further developed, modified and applied in other sectors of the public sector, as well as to commercial organisations.

This brief examination of corporate real estate performance measurement has been selective. However, the theories and ideas considered provide insight into the nature of the measures developed and proposed, and the growing belief that a breadth of measures is required to assess corporate real estate performance — not just one or two. Whether these measures should take the form of the balanced scorecard approach, or a modification of it, or some other formulation or model, is open to further debate. However, regardless of the particular model, the measures included would appear to range from the relatively straightforward cost and revenue based measures (the "hard' measures), to the more qualitative and judgmental based measures (the "soft" measures). This is an aspect of corporate real estate performance measurement that is expected to remain the focus of ongoing interest and research activity for some time to come.

While one aspect of corporate real estate performance measurement is the measurement of the performance of the portfolio and properties within it, a further dimension is the performance measurement of the management function. This is an aspect that is considered in the next section.

Corporate real estate management performance

Studies into the performance of the corporate real estate management function cover a range of aspects that are closely interrelated, namely:

- the level of activity in relation to the corporate real estate management function
- improving the performance of the corporate real estate management function
- assessing the performance of the corporate real estate management function. This includes the identification of activities and characteristics that constitute effective corporate real estate management and the identification of factors that influence management performance.

Level of corporate real estate management activity

The degree of attention or neglect that is accorded to corporate real estate is an indicator of the level of corporate real estate management activity. A number of studies have investigated this aspect and have drawn the conclusion that corporate real estate has traditionally been a neglected resource, albeit more recent studies have suggested that there has been a change in attitude. In one such study Veale (1988 and 1989) concluded that survey results showed a strong correlation between management attitude and effective management of corporate real estate; a view consistent with the findings of other studies.

To conclude that corporate real estate is neglected or undermanaged requires that there exist some standard or framework against which to assess corporate real estate management practice. From the many studies undertaken (for example see Zeckhauser and Silverman, 1981 and 1983; Avis, Gibson and Watts, 1989; Avis and Gibson 1995; Veale 1988 and 1989; Gale and Case 1989; and Schaefers 1999), a number of features or characteristics that are considered to represent or contribute towards active corporate real estate management practice (in other words activities suggesting that corporate real estate is being given a sufficient level of management attention) have emerged. These include:

- existence of a dedicated corporate real estate management unit
- clearly defined estates strategy, contributing to corporate strategy
- reporting of corporate real estate matters at a high level in organisations
- comprehensive inventory of real estate and operation of a management information system
- accounting for real estate on a "property-by-property" basis
- independent evaluation of real estate performance
- charging internal rents.

Following on from the recognition of under management or management neglect of corporate real estate, a number of studies have attempted to examine how management can be developed and improved, and to identify key factors that contribute towards the effective management of corporate real estate.

Improving the performance of the corporate real estate management function

Zeckhauser and Silverman (1981 and 1983) proposed the following seven steps to achieve more effective use of corporate real estate; these same steps were subsequently adopted by Silverman (1987) as the basic structure for his text *Corporate Real Estate Handbook*:

- determine property assets
- set clear, achievable goals
- select appropriate activities
- reorganise the real estate group
- clarify responsibilities
- choose consultants carefully
- set-up a real estate information system.

Bon (1994) provides a basic and brief distillation of corporate real estate management into 10 central principles or points, in which the need for an integrated and coherent property strategy, the alignment of business and property strategies, coherent monitoring of property performance, are emphasised.

From the analysis of these and other studies, similarities can be seen between the suggestions for improving corporate real estate management performance, and the features that if neglected are considered to represent neglect or under management of corporate real estate. Amongst the aspects of particular significance are:

- the preparation of a corporate real estate strategy and its alignment with corporate strategy
- involvement of corporate real estate management at a higher (strategic) level in the organisation
- a good system for management information
- performance measurement and benchmarking of corporate real estate
- the need to develop relationships both internally and externally
- the need to provide a good reliable real estate service and promote it within the organisation.

Further similarities exist between the criteria used for assessing corporate real estate management performance and the features that constitute management neglect, and the recommendations for improving corporate real estate management performance.

Assessing the performance of the corporate real estate management function

In this section the activities and characteristics that are considered to constitute effective corporate real estate management, proposed measures of corporate real estate management effectiveness, and the factors that influence corporate real estate management performance, are considered.

An underlying feature of studies addressing the measurement and analysis of corporate real estate management performance is the reliance on the perceptions of corporate real estate executives and/or senior executives in organisations. For example see Veale (1988 and 1989); Pittman and Parker (1989); Teoh (1993); Housley (1997); and Schaefers (1999). Such analysis has generally been in the form of respondents' perceptions of importance and performance of a range of characteristics and factors. The observation by Pittman and Parker (1989) that objective measures of management performance are very difficult to construct perhaps explains this approach.

The concept of critical success factors has been adopted in a number of studies of corporate real estate management practice, for example Teoh (1993), and Schaefers (1999). Probably the most widely accepted definition of the term critical success factors is that by Rockart (in Isakson and Sircar, 1990), namely:

> Critical success factors thus are, for any business, the limited number of areas in which results, if they are satisfactory, will ensure successful competitive performance for the organisation. They are the few key areas where 'things must go right' for the business to flourish...

Assessment of corporate real estate management performance by reference to critical success factors has generally been carried out by analysis of corporate real estate managers' perceptions of performance. Similarly, assessment of corporate real estate management activity levels has generally been undertaken by reference to the perceptions of corporate real estate managers of the importance attached to the same or a similar set of critical success factors. In some cases attempts have been made to contrast the perceptions of corporate real estate managers with senior managers in the same organisation; for example see Housley (1997). Reliance on perceptions as a basis for assessing the importance and performance of corporate real estate management issues is a weakness of these studies because of the potential for subjectivity and bias. Clearly it is desirable to attempt to test the validity and reliability of such an approach to performance measurement, and to seek to establish objective based measures. However, as indicated by Pittman and Parker (1989), a comprehensive measure of corporate real estate management effectiveness requires that both bottom-line performance and the extent to which the real estate and facilities needs of the organisation are met, are assessed — a task that is regarded as difficult.

From examination of studies that focus on the assessment of corporate real estate management performance, it is evident that there is substantial consensus in terms of the factors that are regarded to be critical to the success of a proactive or high performing corporate real estate management unit; Schaefers (1999) provides the most comprehensive synthesis of these factors. The table below lists the factors used.

Key Factors of Corporate Real Estate Management

(i)	Detailed and up-to-date information on real estate.
(ii)	Centralised keeping of real estate data.
(iii)	Integration of both real estate and corporate information systems.
(iv)	Detailed and formal strategic planning for facilities and real estate asset management.
(v)	Bottom-up integration of strategic planning for real estate and business units.
(vi)	Top-down integration of corporate objectives and strategies in real estate planning.
(vii)	Central location of real estate unit in overall organisational structure.
(viii)	Access to top management.
(ix)	Operation of real estate unit as a separate and distinct responsibility centre.
(x)	Positive attitude by top management towards real estate.
(xi)	Centralised real estate authority and responsibility.
(xii)	Internal renting system for real estate space.
(xiii)	Well-defined and regular real estate performance measurement.
(xiv)	Well-defined and regular strategic real estate control.
(xv)	Transparency of real estate costs.
(xvi)	Professionally trained and qualified human resources in real estate.

Source: Schaefers (1999)

As has been stated earlier, corporate real estate management is a function that has developed substantially during the last 20 years, and continues to do so. Corporate real estate managers are faced with a dynamic environment in which economic, social, technological and working practice changes are affecting the nature of the corporate real estate requirements of organisations. Not surprisingly, recommendations for improving corporate real estate management performance generally relate directly to the activities that are regarded as constituting effective corporate real estate management practice, which in turn are generally consistent with the features used to assess the level of management activity. Additionally there are also relationships between the performance of the management process, and the performance of corporate real estate itself.

From the above, the interrelationship between levels of management activity, recommendations for improving management performance, factors used in the assessment of performance of the management function will be evident.

Concluding comments and observations

The management of investment and corporate real estate portfolios requires a diversity of common, as well as specialised skills on the part of real estate managers. Equally, the assessment of portfolio performance, be it investment or corporate real estate performance, requires similar as well as distinctive processes and methods.

While the management of investment portfolios has generally been regarded as a process that requires active, proactive, tactical and strategic considerations, until recently the management of corporate portfolios has generally been regarded as a more custodial, passive and reactive process. However, this is no longer the case; corporate property management has now come of age and the management style is ever more strategic in approach. This has led to the view being expressed that a paradigm shift has, or is, occurring in corporate real estate management, with the management approach becoming more strategic. Corporate real estate managers, like their investment portfolio manager counterparts, are under increasing pressures to ensure that their portfolios are actively and strategically managed, and that the real estate contributes to the success of the business.

This situation has arisen since the 1970s and 1980s, during a period of growing academic and industry led interest and research into both investment and corporate real estate portfolios and performance. Since then, research into both of these categories of real estate portfolios has progressed at considerable pace, albeit developing in different directions. Without doubt, both of these aspects of real estate management will remain the focus for research for the foreseeable future.

References

Alexander, K (1992) "Facilities Management in the New Organization", *Facilities*, Volume 10, Issue 1, pp. 6–9, MCB University Press, Bradford, UK.

Alexander, K (1994) "A Strategy for Facilities Management", *Facilities*, Volume 12, Issue 11, pp.6–10, MCB University Press, Bradford, UK.

Audit Commission for Local Authorities in England and Wales (1987a) *Local Authority Property A Management Handbook*, HMSO, London, UK.

Audit Commission for Local Authorities in England and Wales (1987b) *Local Authority Property A Management Overview*, HMSO, London, UK.

Avis, M; Gibson, V; and Watts, J (1989) *Managing Operational Property Assets*, Reading University, Reading, UK.

Avis, M and Gibson, VA (1995) *Real Estate Resource Management*, GTI, Reading, UK.

Bank Administration Institute (1968) Measuring the investment performance of pension funds for the purpose of interfund comparison, Bank Administration Institute, USA.

Bannock, G and Partners (1994) *Property in the Boardroom: A New Perspective*, Hillier Parker and Graham Bannock and Partners, London, UK.

Bon, R (1992) "Corporate Real Estate Management", *Facilities*, Volume 10, Issue 12, pp. 13–17, MCB University Press, Bradford, UK.

Bon, R (1994) "Ten Principles of Corporate Real Estate Management", *Facilities*, Volume 12, Issue 5, pp. 9–10, MCB University Press, Bradford, UK.

Bon, R, Luck, R (1999) "CREMRU-JCI survey of corporate real estate management practices in Europe and North America: 1993–1998î, *Facilities*, Volume 17, Issue 5/6, pp. 167–176, MCB University Press, Bradford, UK.

Cabinet Office (1985) *Office Accommodation: A Review of Government Accommodation Management A Report to the Prime Minister*, HMSO, London, UK.

Collis, DJ and Montgomery, CA (1995) "Competing on Resources: strategy in the 1990s", *Harvard Business Review*, July–August, Massachusetts, USA.

Debenham Tewson Research (1992) *The Role of Property — Managing Cost and Releasing Value*, June, Debenham Tewson, London, UK.

Department of Health and Social Security (1982) *Underused and Surplus Property in the National Health Service*, HMSO, London, UK.

Duckworth, SL (1993) "Realizing the Strategic Dimension of Corporate Real property through Improved Planning and Control Systems", *The Journal of Real Estate Research*, Volume 8, Number 4, Fall, pp. 495–510, The American Real Estate Society, Athens, Georgia, USA.

Fama, EF (1970) "Efficient capital markets: a review of theory and empirical evidence", *Journal of Finance*, XXV (2) May, pp 383–417, USA.

Gale, J and Case, F (1989) "A Study of Corporate Real Estate Resource Management" *The Journal of Real Estate Research*, Volume 4, Number 3, Fall, pp. 23–34, The American Real Estate Society, Athens, Georgia, USA.

Gibson, VA (1999) "Quality: the Elusive Measure", Conference Paper, in Investment Property Databank (IPD), (1999) Occupiers Property Databank (OPD) Conference on Information and Performance for Corporate Property, Conference, 25 March, Investment Property Databank, London, UK.

Grimshaw, B (1999) "Facilities management: the wider implications of managing change", *Facilities*, Volume 17, Number 1/2, pp. 24–30, MCB University Press, Bradford, UK.

Hamel, G and Prahalad, PK (1994) *Competing for the Future: Breakthrough Strategies for Seizing Control of your Industry and Creating Markets of Tomorrow*, Harvard Business School Press, Massachusetts, USA.

Hargitay, SE (1986) "Setting up a system", presented to conference on Performance measurement in property investment, March 25, Henry Stewart Conference Studies, UK.

HM Treasury, (2004) Promoting more flexible investment in property: a consultation, (March), HM Treasury, London, UK.

Housley, J (1997) "Managing the estate in higher education establishments", *Facilities*, Volume 15, Issue 3/4, pp. 72–83, MCB University Press Bradford, UK.

IPD Occupiers Property Databank in association with GVA Grimley (1999) Estate management statistics project, Report March 99/18, Higher Education Funding Council for England, Bristol, UK.

Isakson, HR and Sircar, S (1990) "The critical success factors approach to corporate real estate asset management", *Real Estate Issues*, Spring/Summer, pp. 26–31, USA.

Joroff, ML; Louargand, M; Lambert, S; and Becker, F (1993) "Strategic Management of the Fifth Resource: Corporate Real Estate", Corporate Real Estate 2000 Series Report Number 49, Industrial Development Research Foundation, Norcross, Atlanta, Georgia, USA.

Kaplan, R and Norton, D (1992) "The Balanced Scorecard — Measures that Drive Performance", *Harvard Business Review*, January–February, pp. 71–79, Boston, Massachusetts, USA.

Kaplan, R and Norton, D, (1993) "Putting the Balanced Scorecard to Work", *Harvard Business Review*, September–October, pp. 134–147, Boston, Massachusetts, USA.

Kincaid, DG (1994) "A Starting-point for Measuring Performance", *Facilities*, Volume 12, Number 3, pp. 24–27, MCB University Press, Bradford, UK.

Manning, C and Roulac, SE (1999) "Corporate Real Estate Research within the Academy", *Journal of Real Estate Research*, Volume 17, Issue 3, pp. 265–280, The American Real Estate Society, USA.

Markowitz, HM (1952) "Portfolio Selection", *Journal of Finance*, 12 (March) pp 77–91.

Markowitz, HM (1959) *Portfolio Selection: Efficient Diversification of Investments*, J Wiley and Sons Inc, New York, USA.

National Audit Office (1988a) *The Crown Estate*, HMSO, London, UK.

National Audit Office (1988b) *Property Services Agency: Management of the Civil Estate*, HMSO, London, UK.

National Audit Office (1994a) *Financial Health of Higher Education Institutions in England*, December, HMSO, London, UK.

National Audit Office (1989) *Home Office: Control and Management of the Metropolitan Police Estate*, HMSO, London, UK.

National Audit Office (1994) *Department of Education: Management of Office Space*, HMSO, London, UK.

National Audit Office (1996) *The Management of Space in Higher Education Institutions in Wales*, HMSO, London, UK.

Nourse, HO and Roulac, SE (1993) "Linking real estate decisions to corporate strategy", *The Journal of Real Estate Research*, Volume 8, Issue 4, (Fall), pp. 475–495, The American Real Estate Society, Athens, Georgia, USA.

O'Mara, MA (1999) *Strategy and Place: Managing Corporate Real Estate and Facilities for Competitive Advantage*, The Free Press, New York, USA.

Pittman, R and Parker, J (1989) "A Survey of Corporate Real Estate Executives on Factors Influencing Corporate Real Estate Performance", *The Journal of Real Estate Research*, (1989) Volume 4, Number 3, Fall, pp. 107–119, The American Real Estate Society, Athens, Georgia, USA.

Procter, A (1999) "Information and Performance Measurement", Conference Paper, in Investment Property Databank (IPD), (1999) Occupiers Property Databank (OPD) Conference on Information and Performance for Corporate Property, Conference, 25 March, Investment Property Databank, London, UK.

Porter, M (1980) *Competitive Strategy: Techniques for Analyzing Industries and Competitors*, Free Press, New York, USA.

Roulac, SE (1999) "Real Estate Value Chain Connection: Tangible and Transparent", *Journal of Real Estate Research*, Volume 17, Issue 3, pp. 387–404, The American Real Estate Society, USA.

Roulac, SE and Manning, C (1999) "Introduction: Corporate Real Estate Research Thought Leadership", *Journal of Real Estate Research*, Volume 17, Issue 3, pp. 259–264, The American Real Estate Society, USA.

Schaefers, W (1999) "Corporate Real Estate Management: Evidence from German Companies", *Journal of Real Estate Research*, Volume 17, Issue 3, pp. 301–320, The American Real Estate Society, USA.

Sharpe, WF (1964) "Capital asset prices: a theory of market equilibrium under conditions of risk", *Journal of Finance*, 19(3) September, pp. 425–442, USA.

Silverman, Robert A (1987) *Corporate real estate handbook*, New York, USA.

Society of Investment Analysts, (1972) *The measurement of portfolio performance for pension funds*, Society of Investment Analysts, UK.

Teoh, WK (1993) "Corporate Real Estate Asset Management: The New Zealand Evidence", *The Journal of Real Estate Research*, Volume 8, Number 4, Fall, pp. 607–623, The American Real Estate Society, Athens, Georgia, USA.

Weatherhead, M (1997) *Real Estate in Corporate Strategy*, Macmillan Press Limited, Basingstoke, UK.

Veale, PR (1988) Managing Corporate Real Estate Assets: A Survey of US Real Estate Executives, In Cooperation with the International Association of Property Executives (NACORE), The Laboratory of Architecture and Planning, Massachusetts Institute of Technology, Cambridge, Massachusetts, USA.

Veale, PR (1989) "Managing Corporate Real Estate Assets: Current Executive Attitudes and Prospects for an Emergent Discipline", *The Journal of Real Estate Research*, Volume 4, Number 3, Fall, pp. 1–22, The American Real Estate Society, Athens, Georgia, USA.

Zeckhauser, S and Silverman, RA (1981) *Corporate Real Estate Asset Management*, Harvard College, Cambridge, Massachusetts, USA.

Zeckhauser, S and Silverman, RA (1983) "Rediscover your company's real estate", *Harvard Business Review*, January–February, pp. 111–117, Harvard College, Cambridge, Massachusetts, USA.

Appendix A
Corporate Strategy

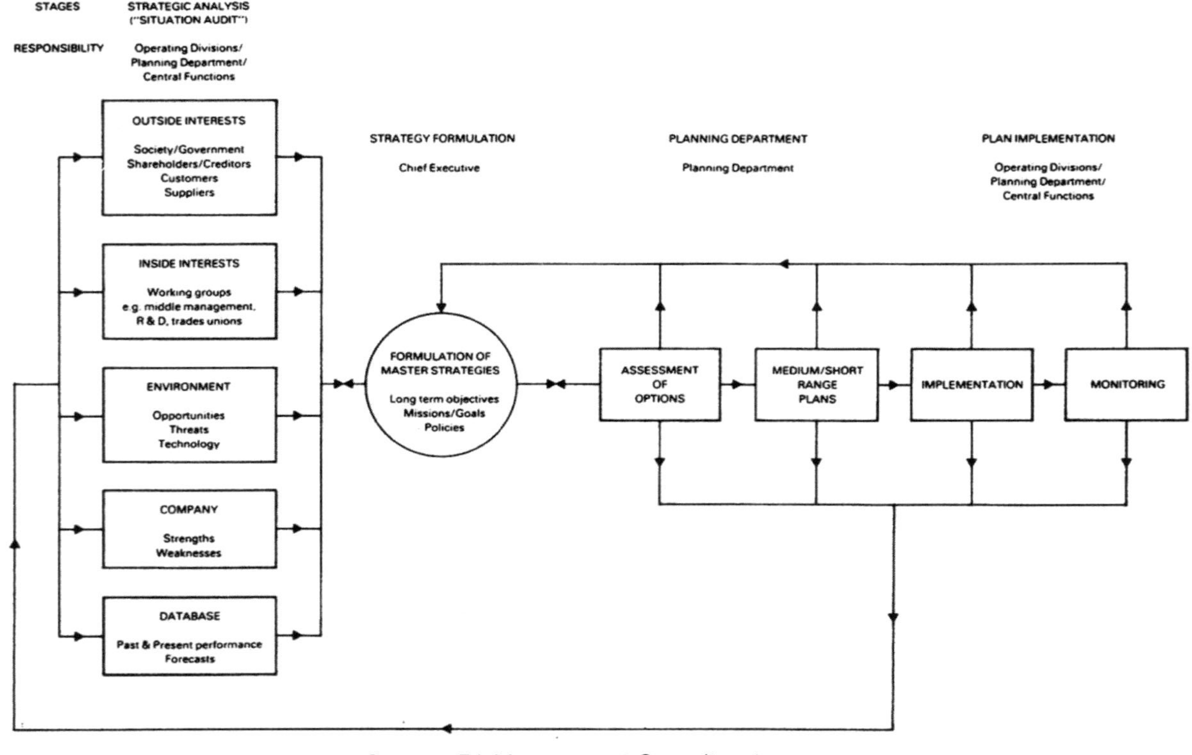

Source: PA Management Consultannts

Appendix B
Strategic Issues for Management Surveyors

Event	*The issues*	*Landlord's perspective*	*Tenant's perspective*	*Comments*
Assignment	Check covenant of assignee Is it allowed in lease Can it be reasonably refused? AGA required?			L&T (Covenants) Act 1995
Part		Can cause complications at renewal	Take to court if unreasonably refused	
Whole		Is proposed assignee's covenant as good	Is proposed assignee's covenant OK?	
Sub-letting	Check covenant of assignee Is it allowed in lease? Is it contracted out of L&T Act 1954? Is it at market rent? See BPF initiative	Probably can refuse if not	Possible reverse premium to subtenant	
Part		Can cause complications at renewal		
Whole		Is it at full market rent?	If over-rented may need reverse premium	*Allied Dunbar* v *Homebase and NCR* cases
Alterations	Detailed drawings and specifications required	Choose to do them and rentalise cost Added value or incumbrance?	Compensation at end of lease?	

Event	*The issues*	*Landlord's perspective*	*Tenant's perspective*	*Comments*
Improvements	Detailed drawings and specifications required			1927 Act
	Landlord & Tenant Act 1927	Choose to do them and rentalise cost	Compensation at end of lease?	
Change of user	Can it be unreasonably withheld?			Illegal to charge a premium
Rent review	Hypothetical terms	Chance to renegotiate terms?	Chance to renegotiate terms?	
	Notice time	Is market rising or falling?		
Arbitrator or expert?		Expert better in a falling market	Arbitrator better in falling market	
Lease termination/renewal				
Options break/review	Conditions to be satisfied prior to exercise	Could negotiate new terms	Could negotiate new terms	
New lease	Agree terms and then check rent			
Repairs/renewals/replacement				
Dilapidations	1927 Act	Do it and recover cost under lease?	Relief under 1938 Act?	

Event	*The issues*	*Landlord's perspective*	*Tenant's perspective*	*Comments*
Insurance				
Bases	Equivalent reinstatement or replacement?			
Loss of rent	2 years or 3 (or more?)		Lease termination if not reinstated?	
VAT	is L/L registered?			
Who insures?	Make sure it is correct	Ask to see documents and receipt	Ask to see documents and receipt	
Environmental issues	Polluter pays	L/L can become liable		
Planning — allow application by tenant?	May require L/L's consent first	Could prejudice s landlord's application Tax implications?		
Other legislation — over-ruling lease covenants	Usually relate to statutory obligations			

Appendix C

Flow-chart through Landlord and Tenant Act 1954 — Part II

(Although this flow chart is prior to amendments by Regulatory Reform (Business Tenancies) (England and Wales) Order 2004, it is still generally correct — but refer to Chapter 6 for details of the easing of time limits by latest legislation.)

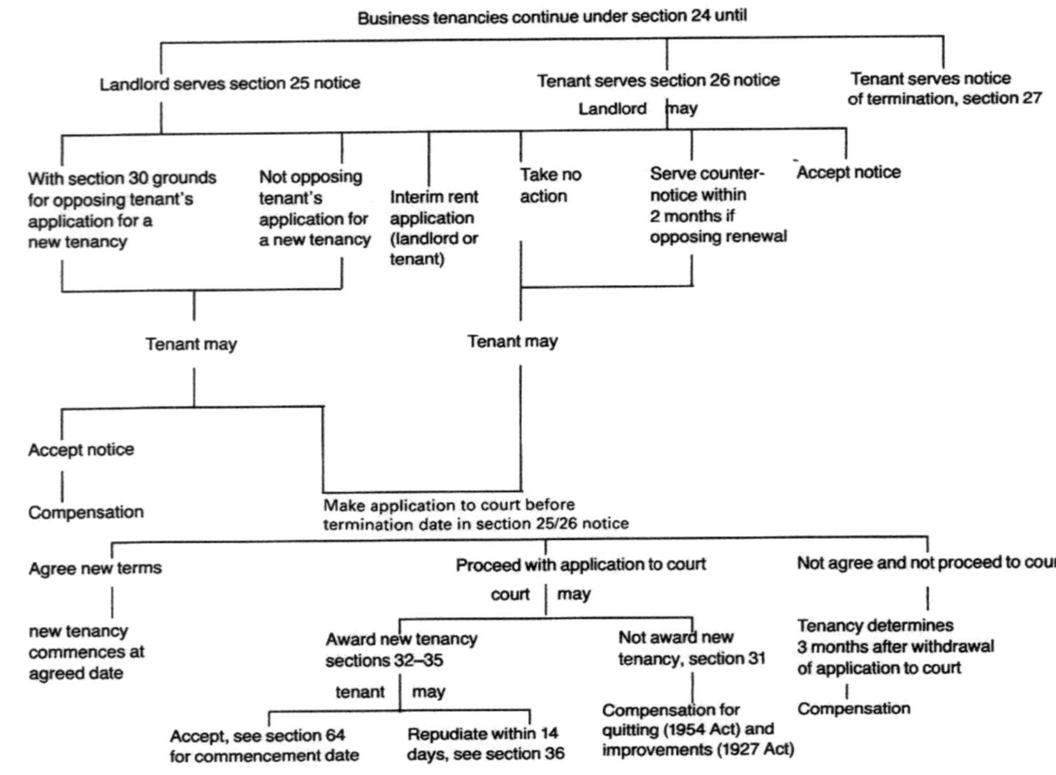
Business tenancies continue under section 24 until
Landlord serves section 25 notice
Tenant serves section 26 notice
Tenant serves notice of termination, section 27
Landlord may
With section 30 grounds for opposing tenant's application for a new tenancy
Not opposing tenant's application for a new tenancy
Interim rent application (landlord or tenant)
Take no action
Serve counter-notice within 2 months if opposing renewal
Accept notice
Tenant may
Tenant may
Accept notice
Compensation
Make application to court before termination date in section 25/26 notice
Agree new terms
Proceed with application to court
Not agree and not proceed to court
new tenancy commences at agreed date
court may
Award new tenancy sections 32–35
Not award new tenancy, section 31
Tenancy determines 3 months after withdrawal of application to court
Compensation
tenant may
Accept, see section 64 for commencement date
Repudiate within 14 days, see section 36
Compensation for quitting (1954 Act) and improvements (1927 Act)

Appendix D

A Code of Practice for Commercial Leases in England and Wales

Second Edition

This updated Code and Explanatory Guide has been produced, at the request of the Department for Transport, Local Government and the Regions, by the Commercial Leases Working Group comprising the Association of British Insurers, Association of Property Bankers, British Retail Consortium, British Property Federation, Confederation of British Industry, Forum of Private Business, Law Society, National Association of Corporate Real Estate Executives (UK chapter), Property Market Reform Group, Royal Institution of Chartered Surveyors and Small Business Bureau. In addition, this code has received support from the British Council for Offices, the British Chambers of Commerce, Council for Licensed Conveyancers and the Federation of Small Businesses.

This Code replaces the First Edition produced by the Commercial Leases Group in December 1995.
ISBN 1-84219-098-9

Property advice

For a selection of local professional property advisers who could represent you call: The Royal Institution of Chartered Surveyors (RICS)Contact Centre on 0870 333 1600.

For a free rent review and lease renewal helpline service for businesses not already professionally represented call the RICS on 020 7334 3806. For the rating helpline call 020 7222 7000. Rating advice also available from the Institute of Revenues Rating and Valuation on 020 7831 3505.

Also free from the RICS: "Rent review — a guide for small businesses" and "Property solutions — A practical guide for your business" available at *www.rics.org*. Send a large stamped, self addressed envelope to Corporate Communications, The Royal Institution of Chartered Surveyors, 12 Great George Street, London SWIP 3AD, or contact the RICS Rent Review and Lease Renewal helpline on 020 7334 3806.

For a "Guide to Good Practice on Service Charges in Commercial Properties", contact the RICS Commercial Property Faculty at the address above (with a large stamped SAE) or the website found at *www. servicechargeguide.co.uk*.

Legal advice

For a free Guide to the Landlord and Tenant Act 1954 write to the Department for Transport, Local Government and the Regions at Eland House, Bressenden Place, London SW1E 5DU.

For information on local solicitors who could represent you, call The Law Society on 020 7242 1222.

For information on local licensed conveyancers who could represent you, call The Council for Licensed Conveyancers on 01245 349599.

Property owners

The trade association which looks after the interests of property owners is: The British Property Federation, 1 Warwick Row, 7th Floor, London SW1E 5ER, Tel: 020 7828 0111 Fax: 020 7834 3442.

Occupiers

Several trade associations look after the interests of occupiers, including the British Retail Consortium and the Property Market Reform Group. The BRC can be contacted on 020 7854 8900. Contact details for the PMRG and other organisations supporting the code may be obtained from the Commercial Leases Working Group Secretariat — (contact details overleaf).

A code of practice for commercial leases in England and Wales

Introduction

This updated Code contains recommendations for landlords and tenants when they negotiate new leases of business premises and where they deal with each other during the term of a lease. The Code consists of twenty-three recommendations which an industry-wide working party, including landlord and tenant representatives, consider reflect current "best practice" for landlords and tenants negotiating a business tenancy. Explanatory guidance notes, set out on pages 4 to 11, provide the background to each of the recommendations.

Landlords and tenants should have regard to the recommendations of this Code when they negotiate lease renewals. Under current legislation if a court has to fix terms for a new lease it may decide not to change the terms from those in the existing lease.

Negotiating a business tenancy (lease)

- **Recommendation 1**: Renting premises: Both landlords and tenants should negotiate the terms of a lease openly, constructively and considering each other's views.
- **Recommendation 2**: Obtaining professional advice: Parties intending to enter into leases should seek early advice from property professionals or lawyers.
- **Recommendation 3**: Financial matters: Landlords should provide estimates of any service charges and other outgoings in addition to the rent. Parties should be open about their financial standing to each other, on the understanding that information provided will be kept confidential unless already publicly available or there is proper need for disclosure. The terms on which any cash deposit is to be held should be agreed and documented.
- **Recommendation 4**: Duration of lease: Landlords should consider offering tenants a choice of length of term, including break clauses where appropriate and with or without the protection of the Landlord and Tenant Act 1954. Those funding property should make every effort to avoid imposing restrictions on the length of lease that landlords, developers and/or investors may offer.
- **Recommendation 5**: Rent and value added tax: Where alternative lease terms are offered, different rents should be appropriately

priced for each set of terms. The landlord should disclose the VAT status of the property and the tenant should take professional advice as to whether any VAT charged on rent and other charges is recoverable.

- **Recommendation 6**: Rent Review: The basis of rent review should generally be to open market rent. Wherever possible, landlords should offer alternatives which are priced on a risk-adjusted basis, including alternatives to upwards only rent reviews; these might include up / down reviews to open market rent with a minimum of the initial rent, or another basis such as annual indexation. Those funding property should make every effort to avoid imposing restrictions on the type of rent review that landlords, developers and / or investors may offer.
- **Recommendation 7**: Repairs and services: The tenant's repairing obligations, and any repair costs included in service charges, should be appropriate to the length of the term and the condition and age of the property at the start of the lease. Where appropriate the landlord should consider appropriately priced alternatives to full repairing terms.
- **Recommendation 8**: Insurance: Where the landlord is responsible for insuring the property, the policy terms should be competitive. The tenant of an entire building should, in appropriate cases, be given the opportunity to influence the choice of insurer.

  If the premises are so damaged by an uninsured risk as to prevent occupation, the tenant should be allowed to terminate the lease unless the landlord agrees to rebuild at his own cost.
- **Recommendation 9**: Assigning and subletting: Unless the particular circumstances of the letting justify greater control, the only restriction on assignment of the whole premises should be obtaining the landlord's consent which is not to be unreasonably withheld. Landlords are urged to consider requiring Authorised Guarantee Agreements only where the assignee is of lower financial standing than the assignor at the date of the assignment.
- **Recommendation 10**: Alterations and changes of use: Landlord's control over alterations and changes of use should not be more restrictive than is necessary to protect the value of the premises and any adjoining or neighbouring premises of the landlord. At the end of the lease the tenant should not be required to remove and make good permitted alterations unless this is reasonably required.

Conduct during a lease

- **Recommendation 11**: Ongoing relationship: Landlords and tenants should deal with each other constructively, courteously, openly and honestly throughout the term of the lease and carry out their respective obligations fully and on time. If either party faces a difficulty in carrying out any obligations under the lease, the other should be told without undue delay so that the possibility of agreement on how to deal with the problem may be explored. When either party proposes to take any action which is likely to have significant consequences for the other, the party proposing the action, when it becomes appropriate to do so, should notify the other without undue delay.
- **Recommendation 12**: Request for consents: When seeking a consent from the landlord, the tenant should supply full information about his/her proposal. The landlord should respond without undue delay and should where practicable give the tenant an estimate of the costs that the tenant will have to pay. The landlord should ensure that the request is passed promptly to any superior landlord or mortgagee whose agreement is needed and should give details to the tenant so that any problems can be speedily resolved.
- **Recommendation 13**: Rent review negotiation: Landlords and tenants should ensure that they understand the basis upon which rent may be reviewed and the procedure to be followed, including the existence of any strict time limits which could create pitfalls. They should obtain professional advice on these matters well before the review date and also immediately upon receiving (and before responding to)any notice or correspondence on the matter from the other party or his/her agent.
- **Recommendation 14**: Insurance: Where the landlord has arranged insurance, the terms should be made known to the tenant and any interest of the tenant covered by the policy. Any material change in the insurance should be notified to the tenant. Tenants should consider taking out their own insurance against loss or damage to contents and their business (loss of profits etc.) and any other risks not covered by the landlord's policy.
- **Recommendation 15**: Varying the lease — effect on guarantors: Landlords and tenants should seek the agreement of any guarantors to proposed material changes to the terms of the lease, or even minor changes which could increase the guarantor's liability.

- **Recommendation 16**: Holding former tenants and their guarantors liable: When previous tenants or their guarantors are liable to a landlord for defaults by the current tenant, landlords should notify them before the current tenant accumulates excessive liabilities. All defaults should be handled with speed and landlords should seek to assist the tenant and guarantor in minimising losses. An assignor who wishes to remain informed of the outcome of rent reviews should keep in touch with the landlord and the landlord should provide the information. Assignors should take professional advice on what methods are open to them to minimise their losses caused by defaults by the current occupier.
- **Recommendation 17**: Release of landlord on sale of property: Landlords who sell their interest in premises should take legal advice about ending their ongoing liability under the relevant leases.
- **Recommendation 18**: Repairs: Tenants should take the advice of a property professional about their repairing obligations near the end of the term of the lease and also immediately upon receiving a notice to repair or a schedule of dilapidations.
- **Recommendation 19**: Business Rates: Tenants or other ratepayers should consider if their business rates assessment is correct or whether they need to make an appeal. They should refer to the DTLR Business Rates — a Guide or obtain advice from a rating specialist. The RICS provides a free rating help line service (see inside back-cover)and advice is available also from the Institute of Revenues Rating and Valuation (IRRV).
- **Recommendation 20**: Service charges: Landlords should observe the Guide to Good Practice on Service Charges in Commercial Properties. Tenants should familiarise themselves with that Guide and should take professional advice if they think they are being asked to pay excessive service charges.
- **Recommendation 21**: Dispute Resolution: When disputes arise, the parties should make prompt and reasonable efforts to settle them by agreement. Where disputes cannot be settled by agreement, both sides should always consider speed and economy when selecting a method of dispute resolution. Mediation may be appropriate before embarking on more formal procedures.
- **Recommendation 22**: Repossession by the landlord: Tenants threatened with repossession or whose property has been repossessed will need professional advice if they wish to try to

keep or regain possession. Similarly, landlords should be clear about their rights before attempting to operate a forfeiture clause and may need professional advice.

- **Recommendation 23**: Renewals under the Landlord and Tenant Act 1954: The parties should take professional advice on the Landlord and Tenant Act 1954 and the PACT (Professional Arbitration on Court Terms) scheme at least six months before the end of the term of the lease and also immediately upon receiving any notice under the Act from the other party or their agent. Guidance on the Act can be found in the Department for Transport, Local Government and the Region's "Guide to the Landlord and Tenant Act 1954".

A code of practice for commercial leases in England and Wales — explanatory guide

The Code of Practice for Commercial Leases consists of the Recommendations set out in this Guide. This Guide gives a brief explanation of the background to the recommendations. Further sources of advice and explanation are listed at the end.

Negotiating a business tenancy (lease)

- **Renting premises**

Roughly a third of business premises in the UK are occupied by rent-paying tenants holding a lease (also called a "tenancy") of the premises. Tenants should choose premises suitable for their short to medium term business plans, in respect of size, location, property and the terms of the lease. Premises might be rented, depending on the individual circumstances, either by the owner granting a new lease, by an existing tenant assigning the lease or by an existing tenant granting a sublease. The terms of a lease should reflect the type, location and condition of the property, the needs and status of the parties, and the state of the property market. All the terms in a commercial lease are normally negotiable.

For business reasons, the landlord or the tenant may wish to keep the details of their transaction confidential, but parties should avoid unnecessary secrecy. This will help the availability of market data.

Recommendation 1: Both landlords and tenants should negotiate the terms of a lease openly, constructively and considering each other's views.

- **Obtaining professional advice**

Unless landlords and tenants are fully experienced in these matters they will benefit from the advice of professional property advisers. Each party should be separately advised by independent advisers as the same person should not advise both parties. Tenants should not place reliance on advice offered by a letting agent acting for the landlord.

The main recognised property professionals are chartered surveyors regulated by the Royal Institution of Chartered Surveyors (RICS), solicitors regulated by the Law Society and Licensed Conveyancers regulated by the Council for Licensed Conveyancers. A surveyor can conduct, or assist in, the negotiations and can advise on the terms, including the appropriate level of rent taking into account the other terms of the letting, the location, size and quality of the property, the state of the property market, the level of business rates and other outgoings, and other relevant matters. A building surveyor can advise about the present condition of the property and about any necessary repairs. For lettings of part of a building, this can include advice about the need for major repairs and renewals of the structure or common parts which might increase service charges. A solicitor can negotiate the detailed text of the lease once the main terms have been agreed. Lease documents often run to many pages and there are no standard forms of lease. A solicitor can also check important matters such as town planning and the landlord's ownership of the property.

Recommendation 2: Parties intending to enter into leases should seek early advice from property professionals or lawyers.

- **Financial matters**

The tenant should find out about the total cost of occupying the premises — rent, service charges, insurance, business rates, utility costs etc. — and ensure that they can be afforded within the budget of the business.

As the landlord will wish to assess the tenant's ability to pay those costs, particularly the rent and any service charge, the tenant should provide written references from accountants, trade suppliers and any previous landlord. If the tenant is a limited company, the landlord may also wish to see audited accounts for the last few years "trading. If this information does not exist or fails to show that the tenant has an adequate financial standing, the landlord may refuse to accept that tenant or may require guarantees from financially viable guarantors, covering not only the rent but also all other liabilities under the lease.

The landlord may also require a cash deposit, frequently of three or six months rent. This "rent deposit" will generally be required as security for service charges and the cost of remedying disrepair or other defaults as well as rent. There should be a proper written agreement covering the amount deposited, whether it can vary, who can hold it, how and when it can be paid over to the landlord or returned to the tenant and which party will receive any interest accruing.

The drawing up of commercial leases involves legal costs. The question of payment is a matter for negotiation between the parties. The Costs of Leases Act 1958 provides that, in the absence of agreement, each side pays its own costs.

Recommendation 3: Landlords should provide estimates of any service charges and other outgoings in addition to the rent. Parties should be open about their financial standing to each other, on the understanding that information provided will be kept confidential unless already publicly available or there is proper need for disclosure. The terms on which any cash deposit is to be held should be agreed and documented.

- **Duration of lease**

The length of the letting is called the "term". Leases are commonly granted for three, five, ten or fifteen year terms, but can be for terms of twenty or twenty-five years or more. A lease carries the protection of the Landlord and Tenant Act 1954, unless the parties agree to its exclusion. If the tenant occupies all or part of the premises when the lease ends, the Act enables a tenant to ask the county court to order the landlord to grant a new lease at a market rent. The landlord can refuse to grant a new lease in certain circumstances set out in the Act, for example if the tenant has seriously defaulted under the lease, or if the property is to be redeveloped or used for the landlord's own business. The tenant can ask the county court to examine the landlord's refusal to grant a new lease. In some cases, the tenant may be entitled to be paid compensation if a new lease is refused. If the lease excludes the Act, the tenant will not have the right to seek a new lease through the courts when the term expires.

Leases can contain a provision (break clause) allowing either the landlord or the tenant (or both) to terminate the lease at a specified date without waiting for the term to expire. This may be advantageous to the party who wishes to end the lease early — such as a tenant who wants to vacate without finding an assignee or subtenant, or a landlord who wants to redevelop — but early termination may cause problems and/or loss to the other party.

Recommendation 4: Landlords should consider offering tenants a choice of length of term, including break clauses where appropriate and with or without the protection of the Landlord and Tenant Act 1954. Those funding property should make every effort to avoid imposing restrictions on the length of lease that landlords, developers and/or investors may offer.

- **Rent and value added tax**

The appropriate level of rent will depend upon the state of the property market, the location, type, age, size, character and condition of the premises and the terms on which the lease is to be granted, especially the duration of the lease and the burden of repairing obligations. Rent is usually payable by quarterly instalments in advance; the usual quarter days being 25 March, 24 June, 29 September and 25 December. One quarter of the yearly rent will usually be payable on these dates. This is not invariable. In some cases, particularly for short term lettings, monthly payments might be appropriate.

Value Added Tax (VAT) will be payable on the rent (and on service charges) if the landlord has elected to waive the building's exemption from VAT. If the landlord has not already done this, it could be done at any time during the lease unless the lease forbids it. If this waiver is made, VAT will be payable by the tenant in addition to the rent and service charge. Many tenants will be entitled to recover the VAT through their business VAT returns.

Recommendation 5: Where alternative lease terms are offered, different rents should be appropriately priced for each set of terms. The landlord should disclose the VAT status of the property and the tenant should take professional advice as to whether any VAT charged on rent and other charges is recoverable.

- **Rent review**

For leases over five years, it is usual for the rent to be reviewed at stated intervals. Usually rent is reviewed to open market rent level — the rent that a new tenant would pay if the property was being let in the open market at the time of the review (the most appropriate basis for review). Alternatives include fixed increases or linking the rent to a published index (such as the Index of Retail Prices) or to the annual turnover of the tenant's business at the premises. Reviews to open market rent normally occur every five years whilst rents linked to indices or turnover are commonly recalculated annually. Not all these methods of review are suitable for every tenant or appropriate to every type of property or business.

If the review is on "upwards only" terms, the rent will not reduce at review but will remain at its existing level even if the market rent or index has fallen. Tenants may find that they would have to pay a higher initial rent where the rent review is to be up or down compared with upwards only, as this transfers the risk of downward movements to the landlord. Financers of property require landlords to ensure that rental income will not fall below a particular level and this may restrict a landlord's ability to agree an upwards/downwards basis.

Recommendation 6: The basis of rent review should generally be to open market rent. Wherever possible, landlords should offer alternatives which are priced on a risk-adjusted basis, including alternatives to upwards only rent reviews; these might include up/down reviews to open market rent with a minimum of the initial rent, or another basis such as annual indexation. Those funding property should make every effort to avoid imposing restrictions on the type of rent review that landlords, developers and/or investors may offer.

- **Repairs and services**

Leases generally state which party will be responsible for carrying out, or for meeting the cost of, repairing and maintaining the fabric and services of the property. The degree to which these burdens are placed on the tenant should take into account the initial condition of the premises and the duration of the lease.

A "full repairing" lease makes the tenant of an entire building responsible for all internal and external repairs and redecoration that become necessary during the term. This includes the roof, foundations, main walls and other structural parts, irrespective of whether or not they are in good condition at the start of the lease. A "full repairing" lease for part of a building requires the tenant to maintain and decorate the inside of the premises and to pay, through a service charge, towards the landlord's costs of maintaining and repairing the common parts and structure and providing services such as porterage, lifts, central heating, etc. Such obligations might require the tenant to carry out, or pay towards the cost of, work to remedy an inherent construction defect which becomes apparent during the term.

Alternatives to "full repairing" terms might include limiting the tenant's repairs to the maintenance of the property in its existing condition, excluding certain categories of repair, and the remediation of inherent defects. The scope or amount of any service charge can be limited or there can be a fixed rent which is inclusive of service costs. If the lease refers to the existing condition of the property, it will be in both parties' interests for a schedule of condition (which can be photographic) to be professionally prepared and kept with the lease documents.

Professional advice should be sought when the tenant is required to carry out initial improvements and repairs, as there may be implications for tax and rent review.

Recommendation 7: The tenant's repairing obligations, and any repairs costs included in service charges, should be appropriate to the length of the term and the condition and age of the property at the start of the lease. Where appropriate the landlord should consider appropriately priced alternatives to full repairing terms.

- **Insurance**

It is usual for the landlord to insure the building and require the tenants to pay the premiums. In the case of multi-occupied buildings, each tenant would be expected to contribute towards the total insurance premium; this may be included in the service charge or may be charged separately. Leases may give the landlord discretion to choose the insurer. Alternatives include allowing the tenant to influence the selection of the insurer (if their lease covers the entire building), or providing that the landlord must arrange the insurance on competitive rates.

The lease should contain provisions covering the situation where there is damage by an uninsured risk or where there is a large excess. These risks vary from time to time and might include terrorist damage. If suitable provisions are not included in the lease the tenant might have to meet the cost of rebuilding in that situation. Alternatives include allowing the tenant to terminate the lease following uninsured damage, although it may be appropriate to allow the landlord to choose to rebuild at his own cost in order to keep the lease in force.

Recommendation 8: Where the landlord is responsible for insuring the property, the policy terms should be competitive. The tenant of an entire building should, in appropriate cases, be given the opportunity to influence the choice of insurer. If the premises are so damaged by an uninsured risk as to prevent occupation, the tenant should be allowed to terminate the lease unless the landlord agrees to rebuild at his own cost.

- **Assigning and subletting**

There are two ways in which the tenant may pass on the lease obligations to a third party; one is by assignment (selling, giving away or paying someone to take over, the lease) and the other is by subletting (remaining as tenant of the lease with the lease obligations but granting a sublease to another tenant who undertakes the same or similar obligations). Leases generally control assignment and subletting. Most require the tenant to obtain the landlord's consent (which cannot be unreasonably withheld) but some leases completely prohibit certain acts such as subletting part of the premises. A new lease, and an existing lease granted since 1995, may expand the landlord's right to control assignments by imposing credit ratings or other financial criteria for assignees. It may also require the assigning tenant to stand as guarantor for any assignee by giving the landlord an "Authorised Guarantee Agreement"; alternatives include giving this guarantee only if it is reasonably required by the landlord, such as where the assignee is of lower financial standing than the assigning tenant.

Recommendation 9: Unless the particular circumstances of the letting justify greater control, the only restriction on assignment of the whole premises should be obtaining the landlord's consent which is not to be unreasonably withheld. Landlords are urged to consider requiring Authorised Guarantee Agreements only where the assignee is of lower financial standing than the assignor at the date of the assignment.

- **Alterations and changes of use**

Leases generally restrict the tenant's freedom to make alterations and often impose tighter control over external and structural alterations than over internal non-structural alterations or partitioning. The lease may absolutely prohibit the work. Alternatives may require the landlord's consent which must not be unreasonably withheld, or may permit the particular type of alteration without consent. The lease may entitle the landlord to require the tenant to reinstate the premises (remove alterations) at the end of the lease; or alternatively reinstatement need only take place if it is reasonable for the landlord to require it.

The permitted use of the premises may be very narrowly defined or there may be a wide class of use. Consent for changes of use can be at the landlord's discretion or, alternatively, the lease may provide that consent is not to be unreasonably withheld. If the provisions of the lease are very restrictive this can hinder the assignment of the lease or the subletting of the property to a different business.

Recommendation 10: Landlord's control over alterations and changes of use should not be more restrictive than is necessary to protect the value of the premises and any adjoining or neighbouring premises of the landlord. At the end of the lease the tenant should not be required to remove and make good permitted alterations unless this is reasonably required.

Conduct during a lease

- **Ongoing relationship**

The relationship between landlord and tenant will continue after the lease has been signed; for example, there may be rent review negotiations or discussions about varying the terms. The landlord may be contemplating planning applications, redevelopment, improvements or making changes in the provision of services.

Recommendation 11: Landlords and tenants should deal with each other constructively, courteously, openly and honestly throughout the term of the lease and carry out their respective obligations fully and on time. If either party faces a difficulty in carrying out any obligations under the lease, the other should be told without undue delay so that the possibility of agreement on how to deal with the problem may be explored. When either party proposes to take any action which is likely to have significant consequences for the other, the party proposing the action, when it becomes appropriate to do so, should notify the other without undue delay.

- **Request for consents**

There may be occasions when the tenant seeks a consent (licence) from the landlord, when for example, the tenant proposes to assign the lease, grant a sublease, change the use of the property, make alterations or display signs. The effect on the landlord will vary with the exact details. In some cases, the landlord will have to pass the request to a superior landlord or to a mortgagee. Most leases require the tenant to pay any costs incurred by the landlord in dealing with such an application.

Recommendation 12: When seeking a consent from the landlord, the tenant should supply full information about his/her proposal. The landlord should respond without undue delay and should where practicable give the tenant an estimate of the costs that the tenant will have to pay. The landlord should ensure that the request is passed promptly to any superior landlord or mortgagee whose agreement is needed and should give details to the tenant so that any problems can be speedily resolved.

- **Rent review negotiation**

Many leases contain provisions for the periodic review of rent; these may be highly technical and may lay down procedures and time limits.

Recommendation 13: Landlords and tenants should ensure that they understand the basis upon which rent may be reviewed and the procedure to be followed, including the existence of any strict time limits which could create pitfalls. They should obtain professional advice on these matters well before the review date and also immediately upon receiving (and before responding to)any notice or correspondence on the matter from the other party or his/her agent.

- **Insurance**

Directly or indirectly, the tenant will usually pay the cost of insuring the premises and the lease will state whether the tenant or the landlord has to arrange this. Where the landlord has arranged insurance, the terms should be made known to the tenant and any interest of the tenant covered by the policy.

Sometimes the lease allows the landlord or the tenant to end the lease if the premises are very badly damaged. If damage occurs but is covered by the insurance, there may be important questions about how, why and by whom the insurance money is spent and the parties should take professional advice as soon as the damage occurs.

Recommendation 14: Where the landlord has arranged insurance, the terms should be made known to the tenant and any interest of the tenant covered by the policy. Any material change in the insurance should be notified to the tenant. Tenants should consider taking out their own insurance against loss or damage to contents and their business (loss of profits etc.) and any other risks not covered by the landlord's policy.

- **Varying the lease — effect on guarantors**

A guarantor may not be liable if the terms of the lease are changed without the guarantor's consent. In some cases the variation may release a guarantor from all liability.

Recommendation 15: Landlords and tenants should seek the agreement of any guarantors to any proposed material changes to the terms of the lease, or even minor changes which could increase the guarantor's liability.

- **Holding former tenants and their guarantors liable**

A tenant who assigns a lease may remain liable for a period for any subsequent breach of the lease terms including failure to pay rent. This liability may also apply to a guarantor for the former tenant. Where payment is made to the landlord under this liability, the former tenant may be entitled to take an overriding lease of the property in order to have some control over the current tenants; legal advice can be obtained about these matters. In certain circumstances, insurance against losses following an assignment may be possible. Landlords must notify previous tenants about arrears of rent and service charges within six months of the amount becoming due, in order to make them liable.

Recommendation 16: When previous tenants or their guarantors are liable to a landlord for defaults by the current tenant, landlords should notify them before the current tenant accumulates excessive liabilities. All defaults should be handled with speed and landlords should seek to assist the tenant and guarantor in minimising losses. An assignor who wishes to remain informed of the outcome of rent reviews should keep in touch with the landlord and the landlord should provide the information. Assignors should take professional advice on what methods are open to them to minimise their losses caused by defaults by the current occupier.

- **Release of landlord on sale of property**

A landlord who sells his interest in the building may remain liable to the tenants to perform any obligations in the lease (for example, in repairing or insuring the building) in the event of failure on the part of the new landlord. It is possible, in certain circumstances, for landlords to terminate their obligations on selling the property through provisions in the lease or, in some cases by seeking the agreement of their tenants and, in the event of objection, decision by a county court.

Recommendation 17: Landlords who sell their interest in premises should take legal advice about ending their ongoing liability under the lease.

- **Repairs**

The landlord may be entitled to serve a notice requiring the tenant to undertake repairing obligations which the tenant has failed to carry out. This notice may be served near or at the end of the term or earlier. The list of repairs is called a "schedule of dilapidations". Disagreements about these are not uncommon and the law on repairing obligations is complex.

Recommendation 18: Tenants should take the advice of a property professional about their repairing obligations near the end of the term of the lease and also immediately upon receiving a notice to repair or a schedule of dilapidations.

- **Business rates**

Uniform Business Rates (UBR) are payable to local authorities and are the responsibility of the occupier (the ratepayer) of the property. In certain circumstances the amount payable can be reduced by appealing against the business rates assessment. Ratepayers should be aware time limits apply to certain appeal procedures and advice on these may be obtained from a rating specialist, who is usually a chartered surveyor.

Recommendation 19: Tenants or other ratepayers should consider if their business rates assessment is correct or whether they need to make an appeal. They should refer to the DTLR Business Rates — a Guide or obtain advice from a rating specialist. RICS provides a free rating help line service (see inside back-cover)and advice is available also from the Institute of Revenues Rating and Valuation (IRRV) — (see inside back-cover).

- **Service charges**

Where the lease entitles the landlord to levy a service charge, details of the services covered are usually set out in the lease and it may contain provisions requiring the landlord to act reasonably or economically. Some leases lay down strict time limits for the tenant to query service charges. Several leading property industry and professional bodies have agreed a Guide to Good Practice in relation to service charges which is available free — ((see inside back-cover).

Recommendation 20: Landlords should observe the Guide to Good Practice on Service Charges in Commercial Properties. Tenants should familiarise themselves with that Guide and should take professional advice if they think they are being asked to pay excessive service charges.

- **Dispute resolution**

Disputes between landlords and tenants can be expensive, time-consuming and divisive. If the lease does not state how a particular dispute is to be settled, the parties may have to go to court. Leases often provide for certain types of dispute to be resolved by particular procedures; for example, it is common to provide that a dispute about rent review is to be referred to an independent surveyor acting either as an arbitrator or as an expert. Professional advice should be obtained about any procedures laid down in the lease.

The parties can agree to appoint a mediator to try to resolve a particular dispute even though the lease does not provide for it. The mediator will consult both parties separately and advise them on the strengths or weaknesses of their case and work towards a settlement. Mediators should be able to keep costs down and achieve an outcome within a short timescale; but if mediation fails, delay and cost will have been incurred and the parties still have to resort to the formal procedures of arbitration, expert determination or court proceedings.

Recommendation 21: When disputes arise, the parties should make prompt and reasonable efforts to settle them by agreement. Where disputes cannot be settled by agreement, both sides should always consider speed and economy when selecting a method of dispute resolution. Mediation may be appropriate before embarking on more formal procedures.

- **Repossession by the landlord**

The lease will contain a clause giving the landlord the right ("forfeiture" or "re-entry") to repossess the property if the tenant breaks any obligations under the lease or becomes insolvent. When a landlord seeks repossession under a forfeiture clause, the tenant (or sub-tenant)may be entitled to claim "relief from forfeiture "from a court, i.e. the right to retain the property despite the breach.

Recommendation 22: Tenants threatened with repossession or whose property has been repossessed will need professional advice if they wish to try to keep or regain possession. Similarly, landlords should be clear about their rights before attempting to operate a forfeiture clause and may need professional advice.

- **Renewals under the Landlord and Tenant Act 1954**

Unless it is excluded, this Act may give the tenant a right to renew the lease when it ends (see under Duration of lease). It contains procedures and time limits that must be strictly followed by both landlords and tenants. Disputes under the Act about whether the tenant should be granted a new lease and about its terms are adjudicated by the county court, but the parties may agree to ask the court to refer all or some aspects to be decided by an independent surveyor or solicitor under the Professional Arbitration on Court Terms scheme operated by the RICS and the Law Society.

Recommendation 23: The parties should take professional advice on the Landlord and Tenant Act 1954 and the PACT scheme at least six months before the end of the term of the lease and also immediately upon receiving any notice under the Act from the other party or their agent. Guidance on the Act can be found in the Department for Transport, Local Government and the Regions, "Guide to the Landlord and Tenant Act 1954".

The Secretariat for the Commercial Leases Working Group can be contacted at: Policy Unit, The Royal Institutionof Chartered Surveyors, 12 Great George Street, Parliament Square, London SW1P 3AD, United Kingdom. T +44 (0) 20 7695 1535 F +44 (0)20 7334 3795
www.commercialleasecodeew.co.uk

Appendix E

Framework for Operational Property Management

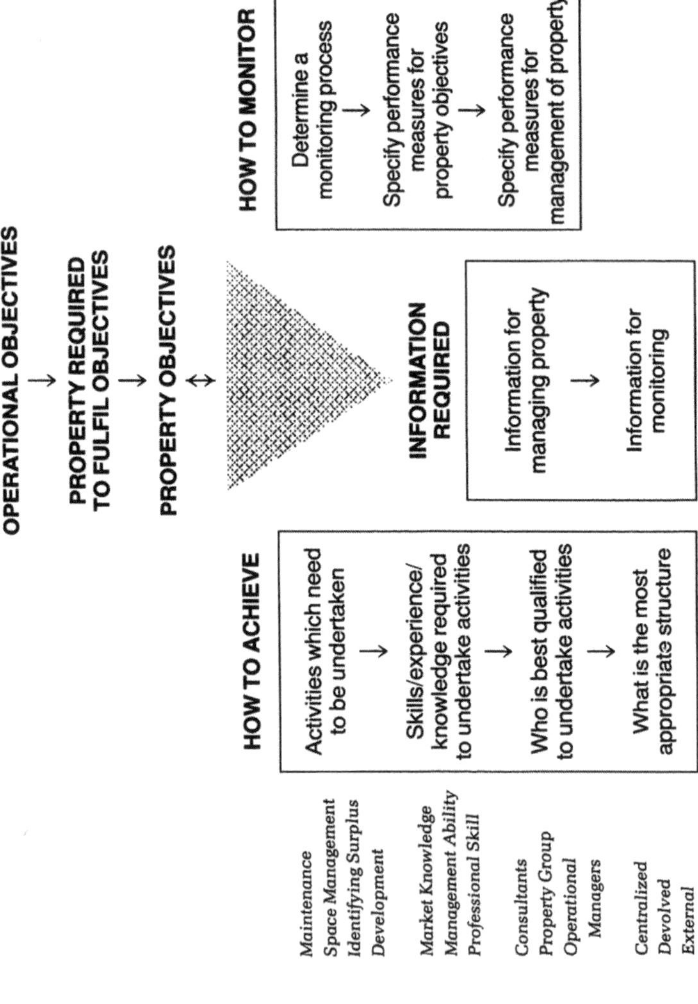
OPERATIONAL OBJECTIVES
PROPERTY REQUIRED TO FULFIL OBJECTIVES
PROPERTY OBJECTIVES
HOW TO ACHIEVE
Activities which need to be undertaken
Skills/experience/knowledge required to undertake activities
Who is best qualified to undertake activities
What is the most appropriate structure
INFORMATION REQUIRED
Information for managing property
Information for monitoring
HOW TO MONITOR
Determine a monitoring process
Specify performance measures for property objectives
Specify performance measures for management of property
Maintenance
Space Management
Identifying Surplus
Development
Market Knowledge
Management Ability
Professional Skill
Consultants
Property Group
Operational Managers
Centralized
Devolved
External

Appendix F

Advisory Guidelines for Commercial Property Management

1 Introduction

1.1 These advisory guidelines set out performance standards for the management of tenanted commercial property.

1.2 The principles set out in the guidelines apply to property managers who undertake property management, either in-house for the landlord, or as managing agents.

1.3 The aim of the guidelines is to encourage property managers to adopt the best practice. This is to ensure that the pre-eminent position of Chartered Surveyors in the commercial management market place remains unchallenged.

1.4 Management is about achieving defined objectives, particularly the provision of quality services which are value for money. This guide is principally concerned about how the person(s) in control of a building should act, particularly how they manage, communicate, and control income and expenditure.

1.5 It should be appreciated that, subject to the tenants' rights under the lease and the law, the landlord has to decide what actions to take or not to take in respect of a property. Equally, the property manager must manage. Nevertheless, effective management needs to take into account what tenants think. Tenants are landlords' customers, and business is generated from satisfied customers. Whilst the objectives of landlord and tenant may differ, mutual benefits will be derived through an understanding of each parties' requirements and by co-operation, resulting in fewer unforeseen disputes for the manager. Effective communication is critical if success is to be achieved.

1.6 A successful property manager must be efficient, effective, and accountable. They should be open about their methods of management, and not be reticent to explain and justify. Consultation and communication will invite challenge, but they are essential if standards are to be improved.

1.7 The guidelines which follow are not drafted to be legally "watertight". For instance, to take the simplest of examples, words such as "regular" will require definition. The task of closing the loopholes should be undertaken when what it is intended to achieve has been settled.

1.8 Unless it is self-evident, each item in the guide is preceded by the rationale for its inclusion. On occasion comments about the implications, or points requiring further consideration, follow the item.

1.9 In deciding what items to include in the guide it is important to bear in mind that the exercise of duties will incur costs and possibly delay. The impact of these will often fall upon the tenants. A pragmatic approach has to be struck about what is reasonable.

1.10 Throughout the guide two further principles have been adopted:

(1) wherever practicable tenants should be given a reasonable time to consider information and proposals put to them, make enquiries, and take advice; and

(2) wherever practicable proposals and decisions by the landlord should be supplied to tenants in writing.

1.11 Existing legislation places some obligation on both landlords and tenants, such as the Offices, Shops and Railway Premises Act. It is intended that this guide should be read in the light of all such legislation. This guide will establish a number of minimum standards of service; delivery will, of course, be dependent upon the contractual relationship between landlord and tenant.

1.12 The property manager should be aware always that it is best practice to endeavour to reach a conclusion to any problem by persuasion rather than litigation.

2 Management duties

2.1 **Disclosure of Landlord**

Tenants should, at all times, be aware of the name and address of the landlord or the landlord's managing agent. It is important to provide the tenants with the name of the property manager, contact addresses and telephone numbers.

2.2 **Management Aims**

The essence of good management is to have clearly defined objectives. The property manager, in conjunction with the landlord, should agree specific objectives for each property, with the manager being pro-active in this respect. Where appropriate, and with the agreement of the landlord, such objectives should be discussed with the tenants. It may be beneficial, for example, to disclose a landlord's intention to redevelop. It might not be appropriate to indicate an intention to sell unless there is a possibility that a tenant could be interested in purchasing. It is clear that in these and other circumstances the management surveyor will be able to provide the correct advice if kept fully briefed.

2.3 **Disclosure of Policies and Related Consultation**

2.3.1 It is good practice to keep the tenants informed of the landlord's objectives in respect of maintenance and management. These will usually be defined with some precision in leases or tenancy agreements, but often not in sufficient detail for tenants to know when action is likely to be taken. Where the costs of management, maintenance, and services are recoverable from the tenants, it is important that they are aware of their potential liabilities and are consulted before the expenditure is incurred.

2.3.2 The property manager should maintain and regularly make available to the tenants a statement of current management policies. This will include ongoing maintenance arrangements, programmes of work planned for the building, the services to be provided, and definite standards of performance which are intended to be achieved.

2.3.3 The property manager should prepare annually a draft budget in respect of proposed expenditure to meet the policy objectives (2.3.2) whether or not this is specifically required by the lease. It should be issued to tenants prior to the beginning of the service charge year allowing sufficient time for the information to be understood. The landlord or agent must consult with the tenants if required by obligations in the lease.

2.3.4 The property manager should communicate effectively with the tenant in order to gain an understanding and acceptance of the service charge budget. Tenants should be able to express their views on the works they would wish to see undertaken, the services to be provided, and on desirable standards. Not only will this be essential in the smooth running of the management, but it will enable tenants to contribute constructively in respect of a cost which is likely to be a significant overhead.

2.4 **Policy Changes in Relation to Recoverable Costs**

Where, without changing legal obligations, the landlord wishes to change or introduce new services, and the cost is recoverable from the tenants, they should be advised prior to implementation. Any such changes must be subject to the terms of the lease.

2.5 **Consultation on Exceptional or Unbudgeted Recoverable Costs**

2.5.1 Tenants should be advised promptly if it appears that the total budget (2.3.3) will be exceeded by more than a prescribed percentage. In any event, interim reports of the progress of actual expenditure against budget will aid effective communication. Also, they should be advised if it becomes necessary to commit expenditure which will increase the tenant's liability in the succeeding accounting period(s) by more than a prescribed percentage in real terms. The lease should be referred to concerning any requirements in respect of varying the collection of the amount of interim charges from the tenants. Wherever possible notification should be given to the tenants prior to any action being taken.

2.5.2 Similarly, tenants should be notified in advance of any exceptional costs, such as for major works. Details of the project, the tender process, the estimated cost, and the contractor to be chosen should be included.

2.6 **Duty to Insure**

2.6.1 To ensure that the landlord is always able to fund restoration of the property and its services in the event of fire or other peril, it is reasonable that the landlord should be required to maintain suitable and adequate insurances. Recovery of the cost from the tenant is normally provided for in the lease.

2.6.2 It is essential for the property manager to make sure that the landlord has insured the property and its facilities against all risks provided for under the lease. Leases vary, but as well as covering the normal comprehensive cover on the buildings the following items should be checked: terrorism,subsidence, loss of rent, third party cover, and other property owners' liability. It is good practice to obtain competitive quotations on a regular basis for the cover required, and to review regularly the sums insured. The tenants should be kept advised of actions taken. Systems should be introduced to notify insurers of any changes of occupation, use, vacant space, and changes in rents to avoid any risks of loss of cover due to non-disclosure of relevant information.

2.7 **Value for Money**

2.7.1 Where costs are recoverable from tenants, the property manager should ensure that the cost of the services provided are good value in terms of the performance standards set. The services should be to a reasonable standard, and the expenditure should be reasonably incurred.

2.7.2 It is good practice to periodically put out to competitive tender the provision of services by contractors approved as being willing, competent and able to provide the required service.

2.7.3 It is recognised that the above would require substantial elaboration to cover such matters as the suitable selection and definition of what is to be the subject of each contract. Also a "de minimus" provision would be appropriate.

2.7.4 Where the expenditure incurred will be charged to the tenants it is good practice that tenders are only invited from contractors associated with the landlord if there is disclosure to the tenants, and they should only be accepted if they are the lowest tenders received.

2.7.5 The RICS Commercial Property Management Skills Panel has produced a paper on service charges which is available from the RICS on application.

2.8 Equitable Distribution of Recoverable Costs/Collection of Funds

2.8.1 In the setting up of a service charge account great care should be given to the apportionment of charges. If it is left to their discretion the property managers should use their professional skill in assessing an equitable basis, including the correct level of sinking funds and on-account charges.

2.8.2 Payments for reserve or sinking funds and for payments in advance should be reasonable in relation to the likely expenditure.

2.8.3 Tenants have the right to challenge an excessive service charge through the Courts.

2.9 Equal Opportunity

The agent must advise the landlord not to discriminate against any tenant(s) or prospective tenant on grounds of sex, race, or marital or religious status.

2.10 Standard of Communication

2.10.1 It is highly desirable that written communications with tenants should, as well as being accurate, be expressed clearly in plain English, concisely and courteously.

2.10.2 Communications must be despatched so that they reach the correct person in the tenant's organisation. The property manager should address all written communications to the tenant at the demised premises or, at the tenant's discretion, such other address as the tenant has provided for the purpose.

2.11 Confidentiality

2.11.1 It may be desirable in the interest of harmony between tenants, but particularly to support their rights to privacy, that contractual, commercial, or personal information relating to the tenants is not disclosed.

2.11.2 Without their consent property managers should not attribute to other tenants or third parties representations made to them by a tenant.

2.12 **Recognition of Representative Bodies**

Good communication, consideration of tenants' views, and cooperation between tenants may be achieved readily and effectively through the recognition by the landlord of bodies representative of some or all of the tenants.

2.13 **Notice of Consultation Meetings**

If a property manager decides to convene a meeting to consult with tenants, reasonable written notice should be given to all tenants who may be affected by the outcome of the matter. It should invite their attendance at a specified place, date and time, and state the matter(s) to be considered.

2.14 **Reasons for Owner's Decisions**

It is reasonable that where the landlord's consent (whether or not it may be reasonably withheld) is required by a tenant and it is refused, or granted subject to a condition(s), the reasons for the refusal or imposition of the condition(s) should be given to the tenant in writing.

2.15 **Enforcement of Covenants (Other than Rental Obligations)**

2.15.1 Tenants are entitled to expect the landlord to take such reasonable steps as are available to ensure that tenants comply with those covenants which benefit another tenant, provided that the landlord is first made aware that breaches are believed to be occurring.

2.15.2 The landlord, having had notice of an alleged breach of covenant the remedy of which would benefit another tenant(s), should promptly, and as reasonably necessary, investigate the allegation. If satisfied that it exists all necessary steps should be taken to bring it to an end, and obtain compensation for restoration of any damage and other expenditure which will otherwise fall to be charged to the service charge account.

2.16 **Maintenance and Investment of Service Charge Accounts**

2.16.1 The property manager should maintain proper accounts of all income and expenditure to and from the service charge account.

2.16.2 Service charge accounts should be prepared promptly after year end. Audited accounts may be required if provided for in the lease.

2.16.3 An agent must comply fully with the latest edition of the RICS Members' Accounts Regulations. In particular, a service charge account must be kept in credit and not funded by the agent.

2.17 **Deposits**

Where a deposit is paid by a tenant or prospective tenant, a receipt must be given which will state the reason why the deposit has been paid. If necessary it should refer to an inventory, and state the time and circumstances, if any, when that deposit\vill be repayable. In addition, it should make clear if interest is to accrue, and what the arrangements are for independent, binding arbitration in the event of dispute as to whether all or part of it is due to be repaid. The deposit should be held in a separate approved account until such time as it is no longer repayable to the tenant.

2.18 **Replies to Correspondence**

It is good practice for property managers to acknowledge receipt of and reply promptly to all correspondence from tenants. Similarly to respond promptly to telephone messages where a response to the tenant is requested. It is also good practice to make a file note of all relevant discussions.

2.19 **Terms of Engagement of Managing Agents**

2.19.1 Where the landlord decides to appoint an agent to carry out any of the management responsibilities, it is reasonable that the tenants should be made aware of the extent of that agent's responsibilities and delegated authority. The landlord should ensure the agent has direct access to sufficient funds to enable the agent to fulfil efficiently and effectively those responsibilities and any other legal obligations.

2.19.2 Where the landlord appoints an agent to carry out any management responsibilities, the agent must therefore:

(a) supply the tenants at the time of appointment, and any new tenants subsequently, with a copy of the Conditions of Engagement, including a statement of the extent of the agent's authority and the basis of the agent's charges if they are recoverable from the tenant;

(b) ensure that at all times there is direct access to draw upon sufficient funds from which expenditure may be paid to fully fulfil those Conditions of Engagement efficiently and effectively and any other legal obligations which the agent may have in law in respect of the property.

2.19.3 Any agent appointed must carry adequate professional indemnity insurance for each and every property managed.

2.19.4 Fees for providing the property manager's service will be either covered under the terms of the lease, in which case they form the contract, or by agreement with the landlord. If the latter the fees must be set at a level which allows the manager a sensible return on the time taken to manage, and at a level which is fair and reasonable to the landlord.

Normally rent collection is charged as a percentage of the rent collected; an alternative is a fixed fee.

With regard to service charges, it is becoming more common to agree fixed fees, but many leases provide for a percentage of expenditure to be charged.

It is important to agree fees in advance if they are not specified in the lease.

3 Variation of conditions of occupancy

The documentation relating to tenancies or management may be found to be legally defective, deficient, or inequitable in relation to proper management. It should be planned to rectify such problems at the time of renewal of leases or management contracts by careful liaison with both landlord and tenant.

4 Collection of money

4.1 Collection of monies, be it rent or service charge, is the core activity of the property manager. It is essential to have in place procedures to provide for:

(a) demanding correct rents/service charge payments in advance of due date;
(b) reminder systems for late payments, normally triggered by the interest trigger date in lease;
(c) system for recovery via Solicitors/Court (see Commercial Property Faculty Paper on Debt Collection).

4.2 Monies collected should have a protocol for action:

(a) rents are normally paid to clients as soon as possible with statements on a quarterly basis;

(b) service charges should be held in a client interest-bearing account with interest accruing for the benefit of the tenants in accordance with RICS Members Accounts Regulations.

4.3 For further details on service charges see Commercial Property Faculty Paper on Service Charges.

4.4 VAT on buildings is a subject on its own, and the Commercial Property Faculty Paper on VAT.

5 Other considerations

5.1 The property manager has to be aware of a number of areas of commercial property law and other laws which have a bearing on the properties managed. It is not the intention of these guidelines to provide detailed analysis of these but to bring them to the attention of the managers.

5.2 Listed below are a number of areas which require careful consideration:

(a) Licences and Assignments:

Care must be given to obtaining full details of any transaction, and information on any variations of original terms, use, etc., by close liaison with Solicitors. Where applicable the manager will be expected to advise on references and the accounts of new tenants.

(b) Health and Safety:

This is a constantly changing area of legislation. The manager must make himself aware of this in relation to the properties managed, taking care to instruct consultants and contractors to comply.

(c) Fire Precautions:

When taking on new management or carrying out or authorising alterations the manager must make sure that full compliance with fire legislation is undertaken.

(d) Maintenance:

It is in the interest of landlord and tenant alike to prepare and comply with a sensible programme of maintenance. Sometimes this work is undertaken under service charge provisions by the manager. However, it is often for the manager to seek compliance with lease covenants by the tenant.

(e) Pro-active Management:
By forward thinking the manager should always be looking at improving the individual property or portfolio. This is often of benefit to both landlord and tenant.

Note:
This is based upon a document issued by RICS and, although it is no longer published by them, it is still a useful aide-memoire for the commercial property manager.

Index